Mapping Innovation Management Systems

In a time defined by accelerating complexity, where technological breakthroughs—from artificial intelligence to blockchain to the emerging contours of Industry 5.0—are rapidly reshaping economic and societal landscapes, the capacity to innovate is no longer optional. It is foundational. Yet innovation itself, while often romanticized as spontaneous or disruptive, demands structure, intent, and governance. It requires systems.

This book is a timely and relevant contribution to this evolving field. The publication of the International Organization for Standardization (ISO) 56001 standard in December 2024 marked a significant milestone: for the first time, organizations have access to a formal, certifiable framework dedicated specifically to innovation management systems (IMS). This book both contextualizes and operationalizes that framework.

Built around the eight core principles of the ISO 56000 series—ranging from leadership and strategy to value realization, adaptability, systems thinking, and collaboration—this volume translates abstract standards into practical, analytically grounded content. It does so through a structured progression across six chapters and a final annex. Each chapter builds on the ISO 56000 principles to examine not only the theoretical underpinnings of innovation management but also its practical manifestations in organizational life.

Drawing from established literature, standards, and real-world practice, this book provides a reliable foundation for anyone seeking to implement or deepen an innovation management system. It will be of particular value to three audiences: practitioners responsible for building innovation capability within their organizations, students eager to understand the logic and structure of innovation systems, and scholars seeking a coherent reference that bridges academic theory with normative frameworks.

In this sense, this book is not just a guide to ISO 56001. It is a comprehensive, well-informed resource for engaging with innovation management as a strategic discipline—one that is rigorous, relevant, and indispensable in the present era.

Mapping Innovation Management Systems

A Guide to ISO 56001 and Its Principles

Alberto Bettanti and Noemi Scimeca

Routledge
Taylor & Francis Group
A PRODUCTIVITY PRESS BOOK

Designed cover image: Shutterstock

First published 2026
by Routledge
605 Third Avenue, New York, NY 10158

and by Routledge
4 Park Square, Milton Park, Abingdon, Oxon, OX14 4RN

Routledge is an imprint of the Taylor & Francis Group, an informa business

ISBN: 9781041247944 (hbk)
ISBN: 9781041247937 (pbk)
ISBN: 9781003743583 (ebk)

DOI: 10.4324/9781003743583

Typeset in Garamond
by KnowledgeWorks Global Ltd.

Contents

Preface

This book was born out of a dual urgency: the growing need for structured innovation within organizations and the opportunity opened by the recent publication of International Organization for Standardization (ISO) 56001—the first certifiable international standard for innovation management systems. Our goal has been to offer a coherent framework grounded in the ISO principles, while also weaving in broader academic thinking, strategic models, and lessons drawn from practice.

This work reflects two different but complementary paths.

Alberto Bettanti brings to this project over two decades of experience at the intersection of consulting, academia, and entrepreneurship. Having worked with global companies and early-stage startups, he has witnessed firsthand the challenges of translating innovation ambition into structured practice. As a professor of management and economics in leading Italian universities, he also contributes to shaping the next generation of business thinkers. And as an entrepreneur, he remains deeply connected to the raw energy of innovation: building ventures, solving problems, and facing uncertainty without filters.

Noemi Scimeca, a recent graduate with strong academic grounding, supported the structuring, research, and drafting process. She contributed to connecting sources, shaping narrative flow, and organizing material in a way that makes complex concepts accessible.

The collaboration between an experienced practitioner and a motivated graduate proved to be more than a working model—it became a reflection of the very subject of this book. Innovation is, at its core, collaborative and multi-generational. It thrives when different lenses—strategic and operational, academic and practical, and senior and junior—interact to challenge assumptions and co-create meaning.

This book is designed for those who want to move beyond slogans and embrace innovation as a system. For leaders and professionals implementing innovation structures within organizations. For students seeking to understand how innovation really works. For scholars who need a robust, integrative resource.

We hope it serves them well.

Alberto Bettanti
Noemi Scimeca

About the Authors

Alberto Bettanti, PhD, has been bridging consulting, academia, and entrepreneurship for 20+ years by helping startups and global enterprises navigate innovation, compliance, and risk management. As the founder of a consultancy, he drives business transformation, integrating innovation strategies, risk frameworks, and International Organization for Standardization (ISO) standards. An academic and researcher, he teaches at top universities, shaping future business leaders in management and economics. And an entrepreneur at heart, he's recently launched startups in beverage and Edtech industries.

Noemi Scimeca is a recent graduate in industrial engineering from the University of Pavia, Italy. She is now pursuing a master's degree in management engineering at the Polytechnic University of Turin. With a humanistic background gained through her linguistic studies, she brings a balanced mix of technical curiosity and open-minded perspective. Her academic journey includes several international experiences focused on language and intercultural learning, fueling her passion for global collaboration.

Introduction

The current landscape of innovation is marked by a convergence of complexity, speed, and systemic change. Organizations today operate within an environment shaped by technological disruption, global interdependence, and rising societal expectations. Developments such as artificial intelligence, blockchain technologies, and the progression toward Industry 5.0 have not only introduced new challenges but also new conditions for value creation. In this context, innovation can no longer be approached as an isolated or occasional activity; it must be treated as a managed process—structured, repeatable, and embedded across organizational systems.

The publication of ISO 56001 in December 2024 represents a significant development in this direction. For the first time, a global standard outlines formal requirements for building and maintaining an innovation management system (IMS). Rather than prescribing a single method or model, ISO 56001 provides a comprehensive framework that supports organizations in developing the capacity to innovate consistently, effectively, and in alignment with strategic objectives.

This book is designed to guide readers through that framework. Drawing from the foundational principles articulated in ISO 56000, the volume interprets and expands on the core elements of ISO 56001 with the aim of offering both conceptual clarity and practical orientation. The structure follows the logic of the standard, progressing through the key dimensions of innovation management while maintaining a focus on real-world applicability.

The first chapter focuses on *leadership*, analyzing how innovation is shaped by visionary direction, cultural commitment, and the strategic role of top management in navigating complexity and enabling transformation. Drawing on both ISO 56000 series requirements and broader theoretical perspectives, the chapter outlines how leadership capabilities underpin the functioning of innovation ecosystems.

The second chapter turns to *innovation strategy*, presenting a comprehensive framework for building long-term competitive advantage. It explores how strategic vision, business model innovation, and the ability to navigate market dynamics form the foundation of a sustainable innovation agenda, while remaining adaptive to shifting external conditions.

The third chapter explores the systematic exploitation of *insights and value realization*, positioning insights as both the cultural and analytical core of innovation. It examines frameworks that help interpret user needs, appropriability mechanisms, and the distribution of innovation-derived value, offering a structured view of how insights are transformed into impact.

The fourth chapter addresses *adaptability and uncertainty management*, emphasizing the importance of experimental approaches, organizational flexibility, and adaptive financial models in dynamic environments. It reflects on how innovation systems must continuously learn and evolve to respond effectively to complexity and risk.

The fifth chapter introduces a *systems approach* to innovation management, providing conceptual and operational frameworks to integrate innovation into organizational processes. The chapter navigates the interaction between technological evolution, performance measurement, and structured innovation processes across different enterprise types.

The sixth chapter explores *organizational culture and collaboration*, analyzing how internal and external relationships, shared values, and network governance mechanisms support innovation. It highlights the relevance of ecosystems thinking, structured partnerships, and cultural transformation in fostering an environment conducive to knowledge integration and co-creation.

An appendix is provided to summarize and contextualize the formal requirements set out in ISO 56001. This section may be especially relevant for organizations preparing to implement or assess their innovation management system in accordance with the standard.

Throughout the book, the most relevant foundational concepts and best practices in innovation management are examined through the interpretive lens of the ISO 56000 Principles. These principles—which include *future-focused leaders, strategic direction, realization of value, exploiting insights, adaptability, managing uncertainty, systems approach*, and *culture*—serve as a unifying thread that connects the practical elements of the IMS with broader theoretical underpinnings. This approach enables a richer, multidimensional understanding of innovation as both a process and a strategic capability.

The content has been developed to serve a range of readers. Practitioners engaged in the design or governance of innovation systems will find a structured reference for aligning their efforts with international standards. Students studying innovation management will encounter a broad and integrated view of the field, with connections to both theory and applied practice. Scholars and educators may also find value in the book's comprehensive treatment of the subject as a resource for teaching and further exploration.

Throughout, the emphasis remains on the systemic nature of innovation management. By examining the interrelations between organizational elements and guiding principles—such as adaptability, insight, value realization, and future orientation—the book aims to clarify how innovation can be managed in a deliberate yet flexible way. The intent is not to offer prescriptive answers but to outline the foundational knowledge and considerations necessary to approach innovation as a strategic capability in complex environments.

In doing so, this book contributes to the evolving body of knowledge around innovation management and offers a coherent framework for those seeking to understand, implement, or advance innovation management systems aligned with ISO 56000 series requirements.

Chapter 1

Leadership for Innovation: Managing Complexity and Driving Transformation

Leadership According to ISO 56000 Series Requirements

Leadership is the backbone of any effective innovation management system. Without it, even the most carefully crafted strategies or advanced tools tend to fall flat. Within the International Organization for Standardization (ISO) 56000 series—recognized globally as a foundational reference for managing innovation—the role of leadership is not only emphasized but also meticulously outlined. This structured guidance offers organizations a roadmap for cultivating forward-thinking leadership practices that can thrive in uncertain, fast-moving environments.

According to *ISO 56000* (ISO, 2020), *Clause 4.3.2.1*, leaders across all organizational levels—whether at the helm of the company or embedded in operational teams—are expected to define and communicate a clear, compelling vision. This vision must do more than just align with business goals; it must motivate individuals to move in unison toward shared innovation outcomes. When well-articulated, this vision acts like a magnetic north, providing purpose, generating momentum, and creating a sense of direction amid the unpredictability of innovation efforts.

But vision alone isn't enough. *ISO 56000, Clause 4.3.2.2*, calls for leaders to foster a climate where challenging the status quo is not only welcomed but encouraged. True innovation leadership means walking a delicate line

DOI: 10.4324/9781003743583-1

between ensuring current operational effectiveness and paving the way for what's next. It's about helping stakeholders feel a sense of ownership in the future, not just keeping them aligned with the present. When this mindset is embraced, organizations are better equipped to anticipate market shifts, explore new technologies, and engage with emerging customer needs before they become urgent imperatives.

The advantages of this forward-leaning leadership approach are far-reaching. As described in *ISO 56000, Clause 4.3.2.3*, the outcomes include increased openness to organizational change, greater cross-functional collaboration, and more precise alignment between emerging innovation opportunities and existing capabilities. It also fosters focused creativity and a disciplined ability to execute promising ideas—two pillars that are often difficult to balance in practice.

Turning to *ISO 56001* (ISO, 2024), the standard drills even deeper into leadership responsibilities. *Clause 5.1.1* specifies that top management must take ownership of both the effectiveness and the efficiency of the innovation management system. This isn't a one-off task—it's a continuous commitment. Leadership must ensure that the system remains consistent with the organization's overall strategic direction and that it's properly embedded in the day-to-day operations. In this view, innovation isn't a separate initiative—it becomes part of the organization's DNA.

This means allocating the right resources, establishing supportive structures, and building internal mechanisms that foster an innovation-ready culture. It also means creating a shared language around innovation, reinforcing the idea that it's not just the domain of research and development (R&D) teams or select departments but a company-wide endeavor.

As also outlined in *Clause 5.1.1*, leaders must emphasize communication. Internal communication is not just about cascading information—it's about clarifying why innovation matters, building shared purpose, and encouraging internal champions. Just as importantly, leaders are expected to commit to ongoing evaluation and performance improvement. The system must deliver results, not just on paper, but in the form of real-world outcomes—new products, processes, services, and ways of working that drive impact.

ISO 56001, Clause 5.1.2, introduces another critical requirement: a leadership mindset that recognizes opportunities as they emerge, especially those grounded in unmet stakeholder needs, whether explicit or latent. This requires a balance: one that weighs potential benefits against risks, demonstrates a healthy appetite for experimentation, and fosters a culture where trial, error, and learning are embraced rather than penalized. A prototyping mindset, combined with the discipline of iteration, becomes a hallmark of resilient innovation systems.

The standard also places strong emphasis on the innovation vision, as described in *Clause 5.1.3*. Leadership must craft a vision that does more than describe incremental gains. It should stretch the organization's imagination, pushing it beyond what it currently believes possible. Such a vision acts as both an internal rallying point and an external signal to partners, customers, and stakeholders. When well-articulated, it enhances the organization's reputation and positions it as an innovator worth watching or working with.

Of course, vision must be paired with strategy. *Clause 5.1.4* highlights the importance of developing an innovation strategy that is coherent, rooted in the organization's context, and clearly aligned with roles, responsibilities, and measurable goals. This strategy cannot be static. It must evolve, based on feedback loops, market signals, and the lessons learned from ongoing innovation initiatives. In dynamic environments, rigidity is a risk. Responsiveness is a competitive advantage.

To further anchor innovation practices within the organization, *Clauses 5.2.1 and 5.2.2* of *ISO 56001* call for the establishment of a comprehensive innovation policy. This policy should not only reflect the organization's core objectives but it must also embody ethical principles and sustainability values. A strong innovation policy acts as a guiding framework, offering clarity about what matters most. Importantly, this policy needs to be broadly communicated, both internally and externally, so that all stakeholders—employees, partners, and customers—understand the organization's commitment and direction.

Finally, *Clause 5.3* reminds us that leadership in innovation is not a passive role. It requires intentional design of systems, strategic alignment of priorities, and relentless attention to what's working and what isn't. Leaders must guide their teams not only through uncertainty but toward purpose, ensuring that innovation delivers measurable value and contributes to long-term competitiveness.

In essence, the ISO 56000 series presents a cohesive framework that reframes leadership not as a single act but as an ongoing commitment. It challenges leaders to weave innovation into the fabric of their organizations—across systems, strategies, and cultures—so that adaptation, growth, and reinvention are not the exception but the expectation.

Leadership and Schumpeterian Dynamics: Driving Creative Destruction

Creative destruction offers a powerful lens to understand how innovation reshapes capitalist economies over time. First introduced by economist Joseph Schumpeter (1883–1950), the concept became central to his broader theory of economic development. Schumpeter (1942), regarded as one of the most influential economic thinkers of the 20th century, used this idea to explain the dynamic, often turbulent forces that drive market evolution and shift competitive landscapes.

In his model, innovation unfolds as a cyclical process—firms constantly push to outpace rivals by introducing what he called "new combinations of means of production." These novel configurations disrupt existing market balances and usher in new competitive realities (Schumpeter, 1911).

Schumpeter's early thinking was shaped by the business environment of early 20th-century Austria, a nation dotted with smaller firms vying for distinction. These businesses competed by innovating, striving to meet customer needs more effectively than their peers. When one succeeded, it often enjoyed a brief window of monopoly, like advantage, pulling market demand in its direction. That success translated into short-term monopoly profits; yet such advantages rarely lasted. Rivals would catch up, imitating and disseminating the successful innovation across the market.

Eventually, this process of diffusion would restore competitive equilibrium, but with a twist: customers ended up with better products, improved services, and lower prices. In this way, innovation becomes both a disruptive force and a driver of progress.

Figure 1.1 captures the essence of how innovation unfolds in waves, one after another continually reshaping the competitive landscape. These waves are driven by new entrants who bring fresh ideas and technologies, creating surges in profitability that, while promising, are inherently short-lived. As competitors catch up and adopt these innovations, the advantage of the early movers start to fade. Yet, the broader market benefits: increased competition leads to better offerings, lower prices, and, ultimately, social progress.

Schumpeter referred to this recurring upheaval as *creative destruction*—the constant dismantling and rebuilding of market structures—and saw it as a vital engine for economic development. In this view, innovation isn't just occasional disruption; it's the central force propelling capitalism forward, with each cycle of transformation pushing the economy into new territory.

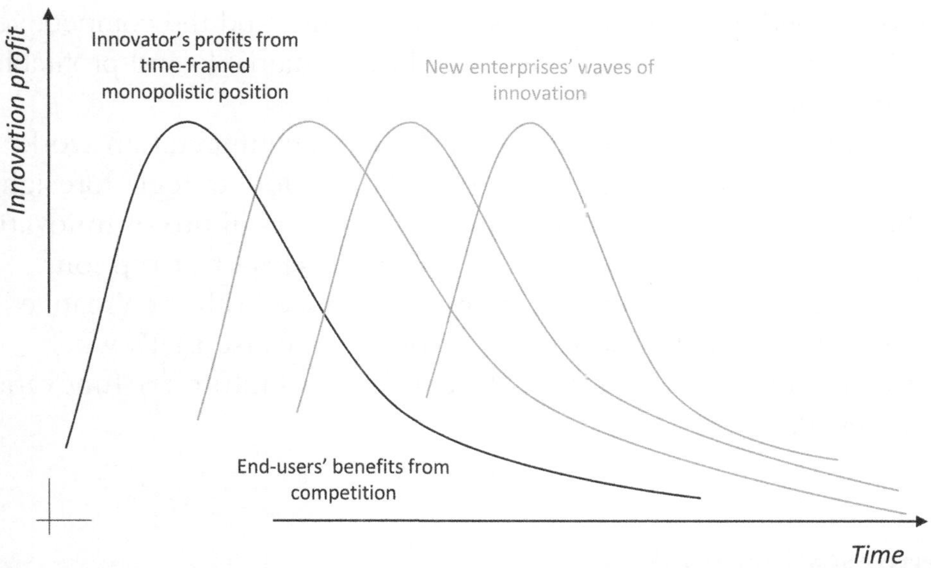

Figure 1.1 Schumpeter's innovation model.

Schumpeter outlined five key forms that these "new combinations" of innovation could take: launching a new product, introducing a novel method of production, venturing into untapped markets, securing new sources of raw materials, or reorganizing industries through structural changes, such as the rise or fall of monopolies. At the center of this process stands the entrepreneur. For Schumpeter, this figure isn't just a businessperson; they are a catalyst, someone with the vision and drive to shake up the status quo.

Although both Schumpeter and Keynes allude to the unique motivations that push such individuals—terms like "entrepreneurial genius" or "animal spirits"—they don't delve deeply into the psychological traits behind them. Still, Schumpeter acknowledges that human creativity and ambition are indispensable forces, working alongside more traditional economic inputs.

An important tension arises when considering how long a firm should enjoy the monopoly created by a successful innovation. Extended periods of market dominance can incentivize companies to invest in bold, long-term ideas. But they may also stifle competition and limit consumer choice. On the flip side, if monopolies are too short-lived, the reward for innovation may not be enough to justify the effort and risk—potentially dampening innovation altogether.

Navigating this trade-off requires thoughtful leadership. Business leaders must evaluate factors like the uniqueness of the innovation, how easily it can be copied, legal protections such as patents, and the competitive strategies at play. These decisions shape how sustainable and profitable innovation can be.

Ultimately, leading through innovation in a Schumpeterian world demands more than technical expertise. It calls for strategic foresight: the ability to seize fleeting opportunities, foster a culture of innovation, and position the organization to thrive amid cycles of disruption. Entrepreneurs and executives alike must embrace both the creative spark and the inevitable destruction it brings because that's where growth truly begins (see section "Establishing a Culture for Innovation" in Chapter 6).

Corporate Leadership

Leadership for Organizational Innovation

Today's large organizations are navigating an increasingly volatile and dynamic innovation landscape. As the pace of technological change and market disruption accelerates, firms are compelled to evolve beyond traditional approaches rooted in incrementalism—models that once excelled at optimizing small-scale improvements but now struggle to generate the kind of radical, game-changing solutions required for competitive advantage. To thrive, companies must rethink the very foundations of how innovation operates within their structures. This transformation isn't merely tactical; it demands a systemic reimagining of internal processes, organizational frameworks, and—perhaps most crucially—the culture and leadership philosophy that shape how people work.

Such a shift involves more than introducing new tools or adopting agile methodologies. It requires a cultural and structural metamorphosis, touching everything from how communication flows across silos to how authority and decision-making are distributed and to how innovation capabilities align with external technological ecosystems. Entrepreneurial thinking must be nurtured throughout the organization, along with the ability to collaborate across functional and hierarchical boundaries. Innovation today is as much about how organizations connect outward—with partners, institutions, or regulatory bodies—as it is about what happens within their walls.

Figure 1.2 helps visualize this complexity by offering a clear comparative framework of internal barriers and enablers. Based on

	Weak Factors	Driving Factors
Strategy	• "Hurdles too high": managers reject Big-I opportunities that at first glance appear too small. • "Scope too narrow": only products count, even though Big-I can come from a range of sources such as process and marketing.	• "Innovation portfolio approach": see an example in Figure 1.3, "A Small-I to Big-I portfolio approach."
Process	• "Strangle innovation with tight controls": same mainstream planning, budgeting, and reviews applied to greenfield.	• "Add flexibility to Big-I innovation system": reserve pools of extra funds for innovation cycles.
Structure	• Small-I vs. Big-I innovation systems: connections too loose and separations too sharp.	• "Facilitate close connections between Big-I and Small-I business": managers from Smal -I business rotating through Big-I groups or advisory boards overseeing innovation projects. • Create a united vision to iEcosystem engagement approach among the different perspectives of corporate units (R&D, Legal, HR, CVC, Business Development, etc.).
Skills	• "Leadership too weak": undervaluing the human side of Big-I, top managers put best technical people in charge instead of the best leaders. • "Communication too poor": communication not effective enough to share the Big-I tacit knowledge.	• Within the Big-I innovation system, select for leadership and interpersonal skills, and surround Big-I environment with a supportive culture of collaboration.

Figure 1.2 Weak organizational and robust driving factors in innovation systems. (From Kanter, R. M., Innovation: The Classic Traps, 2006. With Permission.)

the work of Rosabeth Moss Kanter (2006), the figure maps out weak organizational factors—such as high internal hurdles for radical ideas, narrow product scopes, rigid budgeting controls, and siloed communication channels—against robust driving forces like innovation portfolio strategies, dynamic resourcing, cross-functional integration, and leadership empowerment. This dual perspective enables leaders to assess their organization's current state and identify leverage points for strategic transformation.

Importantly, the table emphasizes that innovation is rarely stifled by a single failure point. Rather, it falters when multiple weak factors converge, compounding internal resistance and eroding momentum. Conversely, activating a handful of driving forces—such as enabling fluid collaboration between incremental and radical innovation groups or appointing boundary-spanning leaders to champion projects—can generate powerful ripple effects. The value of Figure 1.2 lies not only in diagnosing bottlenecks but also in illuminating a path toward a more adaptive, responsive, and entrepreneurial innovation system.

Figure 1.2 reveals that overcoming these barriers demands purposeful action: broadening the innovation portfolio beyond conventional boundaries, relaxing overly rigid governance systems, building connective tissue between small-scale and large-scale innovation efforts, and cultivating leadership skills that enable trust, communication, and adaptive learning. Rather than treating innovation as an isolated function, leading firms embed it into the organizational fabric.

This is where the concept of "change platforms," as introduced by Hamel and Zanini (2014), becomes instrumental. Traditional change programs—anchored in top-down planning, heavy documentation, and hierarchical controls—are ill-suited to navigate complex, uncertain environments. Change platforms, by contrast, are dynamic and participatory. They invite employees across all levels to co-own transformation by identifying bottlenecks, contributing ideas, and collaborating on prototypes or pilots. These platforms replace command-and-control structures with shared ownership and decentralized experimentation, accelerating learning cycles and diffusing innovation capabilities across the firm.

However, as Corbett (2018) argues, participatory platforms alone are not sufficient. Innovation must be supported by robust institutional scaffolding—formal governance mechanisms, defined roles, repeatable routines, and incentive systems that translate grassroots energy into organizational outcomes. Without these systems in place, even the most promising innovation efforts risk stalling or being absorbed into operational inertia. Corbett emphasizes the need for career pathways for innovators, internal funding models, portfolio oversight, and structured feedback loops: all essential ingredients for turning a culture of innovation into sustained value creation.

From a strategic standpoint, this evolution calls for organizations to move away from overly risky "big bet" models that hinge on one breakthrough project and toward innovation portfolio approaches that spread risk across a diverse mix of initiatives. As illustrated in Figure 1.3, these portfolios vary in scope, novelty, and investment, encompassing everything from incremental improvements in existing technologies to radically novel explorations into uncharted problem spaces. The figure reveals how balancing high-budget, high-risk radical innovation projects (Big-I) with more modest incremental initiatives (Small-I) allows firms to maintain a pipeline of potential solutions while adjusting dynamically to external shifts.

Agility becomes the cornerstone of process redesign. Organizations must be capable of pivoting swiftly in response to new information or market disruptions. This requires not just structural integration between Small-I

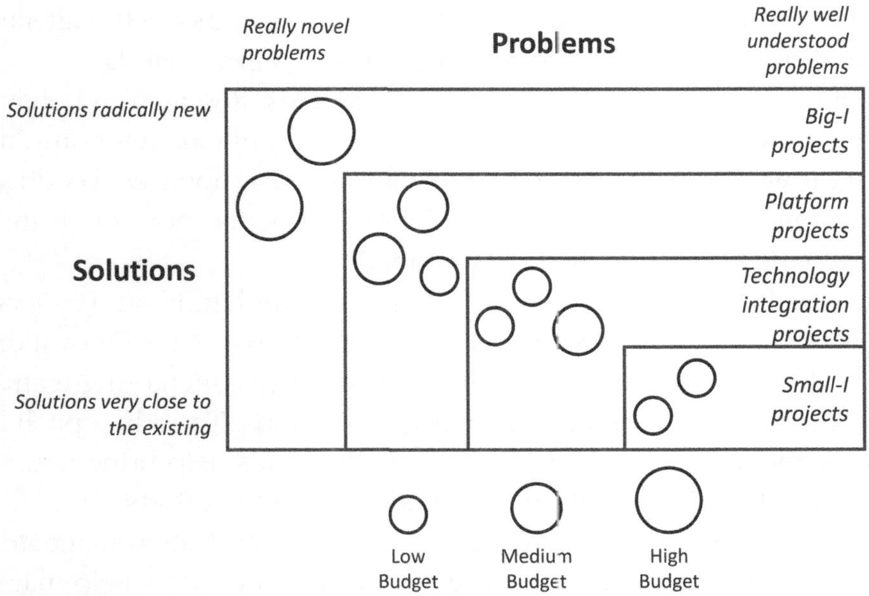

Figure 1.3 A Small-I to Big-I portfolio approach.

and Big-I innovation streams but also strategic partnerships with startups, academic institutions, and corporate venture units. A Chief Innovation Officer (CINO) reporting directly to the CEO can serve as a catalyst here, ensuring that innovation receives visibility, legitimacy, and alignment with core business priorities (Kanter, 2006; Di Fiore, 2014).

Yet even the best-designed systems can fail if the human side is neglected. Leadership and relational skills are often underestimated in their importance to Big-I innovation. Successful teams are those that can tolerate ambiguity, foster trust, share insights fluidly, and operate across boundaries. Organizations must recruit and develop talent capable of operating in these dynamic environments, while resisting the myth of the lone intrapreneur. While stories of visionary mavericks abound, evidence suggests that transformational innovation is almost always the product of institutional scaffolding, not individual heroics. Kodak's failure to commercialize its own digital photography innovations stands as a cautionary tale: without the right structural support, even brilliant ideas can die on the vine.

To embed innovation systemically, companies must formalize governance processes, create dedicated innovation roles, allocate time and resources for exploratory work, and build career pathways for intrapreneurs. Developing entrepreneurial capacity—the ability to move quickly, take smart risks, and learn from failure—is especially vital for organizations hoping to collaborate with external ecosystems. But building this capacity requires thoughtful attention to incentive systems, managerial sponsorship, and organizational slack.

Incentives need to go beyond financial rewards. Research in behavioral economics and motivation theory shows that recognition, autonomy, and meaningful work often carry more weight than cash bonuses. Feedback-rich environments, peer validation, and visibility can be far more motivating, especially when paired with clear accountability.

Managerial support must also be intentional and hands-on. Leaders at all levels—not just senior executives—must create space for innovation to happen. This includes dedicating time and resources, mentoring teams, and embedding innovation into performance evaluations. Tools like pitch days, internal accelerators, and dedicated innovation funds help bring structure to what might otherwise be informal or under-resourced efforts.

Structural enablers are equally important. Clear innovation mandates, rapid decision-making channels, and lean governance frameworks help maintain momentum. And perhaps most importantly, risk-taking must be normalized—not glamorized but understood as an inherent part of progress. When failure is reframed as learning, teams are more likely to experiment and iterate.

Ultimately, sustained innovation is not the product of isolated initiatives, but of deliberate orchestration across strategy, structure, culture, and people. When these elements align—and when organizations treat innovation as a system rather than a series of one-off projects—they position themselves not just to survive disruption but to lead it.

Leadership for Balancing Innovation and Entrepreneurial Capabilities

Today's organizations operate in a landscape shaped by constant disruption, evolving technologies, and increasingly demanding markets. To thrive in this environment, corporate leadership must learn to walk a delicate line, cultivating both deep-rooted innovation capability (I-Cap) and fast-moving entrepreneurial capability (E-Cap). These are not interchangeable strengths. Rather, they represent two fundamentally distinct ways of thinking, organizing, and acting, each with its own rhythms, priorities, and cultural requirements.

As illustrated in Figure 1.4, I-Cap and E-Cap pull from very different toolkits. I-Cap relies on methodical, science-driven development. It is powered by STEM talent, long-term R&D investments, access to advanced infrastructure such as specialized laboratories or simulation environments, and close relationships with highly sophisticated

Dimension	Innovation Capability (I-Cap)	Entrepreneurial Capability (E-Cap)	Leadership's Role
Human Capital	STEM expertise and R&D skills; interdisciplinary collaboration is key.	Entrepreneurial talent with risk tolerance and business acumen.	Foster mentoring, cross-functional teams, and networking to connect technical experts with entrepreneurial innovators.
Funding	Long-term investment in R&D is essential for continuous innovation.	Access to risk capital from investors (e.g., venture capital, angel investors) to scale new ventures.	Secure dedicated R&D budgets and build robust partnerships with external investors.
Infrastructure	Specialized laboratories, state-of-the-art equipment, and advanced technical resources drive effective research efforts.	Flexible office spaces, co-working hubs, and rapid prototyping facilities are critical for agile startup operations.	Invest in both advanced and flexible infrastructures; streamline processes to support swift experimentation and prototyping.
Demand	Stable demand from sophisticated customers requiring complex, tailored solutions validates innovative efforts.	Emerging market opportunities and unmet needs provide fertile ground for startup growth.	Build and nurture strategic relationships with key buyers, enabling real-world validation through feedback loops.
Culture & Incentives	A culture that values intellectual property protection, rewards breakthrough innovations, and encourages disciplined R&D efforts.	A culture that incentivizes risk-taking, agile experimentation, and entrepreneurial problem-solving.	Develop balanced incentive systems that recognize both research excellence and entrepreneurial innovation; align short-term tactics with long-range strategy.

Figure 1.4 I-Cap vs. E-Cap leadership.

customers who validate and shape innovation needs. The culture underpinning I-Cap is one of rigor, protection, and continuity; it rewards patents, values precision, and thrives on structured problem-solving over time.

In contrast, E-Cap runs on a different fuel. It thrives on ambiguity, flexibility, and urgency. E-Cap demands individuals with a high tolerance for risk and a deep instinct for opportunity—people capable of identifying weak signals, launching fast experiments, and pivoting without hesitation. Here, access to external capital (from venture capital or angel investors), flexible infrastructure (such as co-working hubs or maker labs), and rapidly evolving markets are the key enablers. It's a domain where failure is fast, frequent, and valuable—a space built not on protecting innovation but on pressure-testing it through rapid iteration and user feedback.

The question for leaders, then, is not which capability to pursue but how to integrate both.

That's where leadership becomes transformative. The most impactful leaders today don't merely oversee separate innovation or venture functions. They serve as architects of interaction, connecting R&D engineers with startup founders, linking internal science teams to external incubators, and aligning the long-range vision of I-Cap with the nimble experimentation of E-Cap.

As shown in Figure 1.4, this integration requires a wide range of leadership interventions: from securing patient capital for scientific discovery to cultivating strong relationships with external investors and from investing in physical innovation infrastructure to building internal accelerators that mimic the agility of startups. Leaders must enable collaboration not just across departments but across mental models. They must mentor across generations and functions, bridging traditional technologists with entrepreneurial thinkers in ways that generate new synergies.

Just as importantly, leaders must also shape culture. I-Cap thrives when intellectual property rights (IPR) is protected and deep expertise is nurtured. E-Cap, by contrast, blossoms in cultures that reward speed, agility, and risk-taking. The challenge is to design incentives and norms that allow both to coexist without undermining each other. That might mean offering long-term research incentives alongside short-term venture challenges. Or embedding entrepreneurial training within core technical functions so that even the most specialized scientist sees him/herself not just as a researcher but as a potential founder.

This balancing act is not a luxury; it is a necessity. In a world of compressed timelines and blurred industry boundaries, organizations must be able to innovate continuously while also sensing and seizing emergent opportunities. A company with only I-Cap risks becoming brilliant but slow. One with only E-Cap may be fast but fragile. True resilience and long-term value creation lie in the synthesis of both.

Ultimately, this is where leadership makes its mark: not simply in allocating resources but in designing the connective tissue between disciplines, teams, and time horizons. The leaders who succeed in this are not just managing today's complexities; they are building the capabilities for tomorrow's reinvention.

Future-Focused Leadership

Innovation leadership isn't just a desirable trait in modern organizations; it's a vital cornerstone of long-term success. What sets truly innovative leaders apart isn't only their expertise but a blend of mindset, empathy, and behavioral nuance that inspires transformation from within. Decades of research into what makes these leaders effective have revealed a core set of ten behavioral characteristics that repeatedly show up when innovation thrives (Zenger & Folkman, 2014).

At the top of the list is strategic foresight—the ability to see beyond the present and articulate a compelling future. These leaders know how to paint vivid scenarios for their teams, not as distant visions, but as goals worth striving toward. They're also deeply tuned into the needs of their customers. This empathy helps them connect the dots between market demands and bold new ideas, translating insights into tangible innovations.

But innovation doesn't flourish in a vacuum; it needs a culture of trust. Great leaders work to build psychologically safe spaces where people feel free to take risks, speak up, and fail without fear of blame. Mistakes aren't punished; they're mined for insight. This mindset shift—from perfection to progress—transforms teams into agile, creative engines.

Integrity is another defining trait. The best innovation leaders hold true to ethical standards and strategic intent, even when it's not the popular or politically expedient path. They create environments where openness flows both ways where good ideas can come from any level and upward communication is welcomed, not avoided. That openness makes innovation a collective endeavor, not a top-down directive

These leaders are also persuasive, not in a pushy or authoritarian sense, but in how they galvanize support through authenticity and conviction. They inspire buy-in rather than demand compliance. They stretch their teams with ambitious yet achievable goals and break down bureaucratic bottlenecks that slow innovation. Honesty and transparency define how they communicate, making expectations clear and engagement high.

At the heart of these ecosystems, intrapreneurs thrive.

Intrapreneurs—those entrepreneurial minds working within larger organizations—are often the sparks that ignite transformative change. What drives them isn't a paycheck, but the freedom to make a difference. They're observant, quick to seize emerging opportunities, and skilled in adapting to change. One of their secret weapons? "Greenhousing"—the ability to shield delicate, early-stage ideas from premature criticism until they're ready to stand on their own.

These innovation-minded individuals use creative thinking tools like brainstorming and design thinking, bringing in diverse perspectives to shape new solutions. And when the landscape shifts, they pivot, adjusting direction with strategic intent—just as business legends like Steve Jobs or Jeff Bezos did. But it's not just vision that defines them; it's integrity. Their purpose-driven commitment, humility, and self-assurance make them respected agents of change.

Still, none of these sticks without communication.

Strategic communication is the glue that binds innovation leadership. Without it, transformation falters. Leaders must not only set direction but also rally people around it—across departments, up and down hierarchies, and beyond the organizational boundaries. They must share the "why," not just the "what." Good communication challenges old assumptions, unites diverse stakeholders, and spreads the lessons that fuel further innovation.

Importantly, it's not just about data. The most effective leaders balance facts with stories, narratives that resonate emotionally and motivate action. This combination fosters shared purpose and celebrates achievements in a way that energizes the entire organization. It's through this kind of communication that cultures of innovation are born and sustained.

In the end, building innovation leadership is a deliberate journey. It demands more than technical competence; it calls for behavior, vision, and empathy. Organizations that prioritize these leadership capabilities, cultivate entrepreneurial talent, and amplify strategic communication will be better prepared to thrive in fast-moving markets (Govindarajan & Desai, 2013).

Balancing Management and Leadership

Within the realm of organizational leadership, one of the most enduring yet misunderstood dynamics is the relationship between management and leadership. Too often, these two functions are viewed as interchangeable, or worse, as opposing forces. In reality, they are complementary, equally essential elements of a well-functioning organization. Each plays a distinct role in ensuring not only the company's present stability but also its future vitality.

Management is fundamentally about mastering complexity. It operates through structure, systems, and processes designed to bring order to the inherent chaos of business operations. Managers focus on planning workflows, defining procedures, allocating resources, and solving logistical problems as they arise. Their role is to keep the organizational engine running smoothly, to ensure efficiency, reduce variability, and optimize performance. This function is especially crucial for maintaining quality in products and services, enforcing consistency, and delivering steady, incremental improvements. Without it, the organization risks losing its footing.

Leadership, on the other hand, is future-facing. Where management stabilizes, leadership mobilizes. It is concerned with inspiring people, building shared purpose, and crafting a vision that compels others to move toward something greater than the current state. Leaders are the architects of transformation. They don't just ensure performance today; they lay the groundwork for strategic renewal and future competitiveness. Their success rests on two key pillars: maintaining operational excellence in the short term and seeking out opportunities for long-term growth, differentiation, and innovation. These dual responsibilities demand a deep well of emotional intelligence and an ability to ignite motivation at both the individual and collective level. Leaders must model the behaviors they wish to see, nurture curiosity, and cultivate a culture of continuous learning (Goleman, 2000; Kotter, 2001).

This distinction between managing and leading becomes especially critical when we turn our attention to innovation. Managing innovation requires a disciplined approach to execution: tight project controls, efficient resource allocation, and careful monitoring of timelines and risks. But these managerial competencies alone are insufficient. Innovation flourishes only when paired with visionary leadership—leadership that encourages experimentation, tolerates uncertainty, and is willing to pursue unproven paths. The leader's role, in this context, is to be both navigator and explorer: to provide structure without stifling creativity and to champion a culture where new ideas are welcomed rather than feared.

The ideal innovation leader walks this fine line with care. They build environments in which failure is not punished but studied, where experimentation is not chaotic but bounded and deliberate. Within these environments, the emphasis is on learning from setbacks and not simply avoiding them. Organizations that treat every mistake as a data point, every misstep as a source of insight, are those that mature into resilient innovators. Psychological safety becomes essential here. Employees must feel empowered to take risks, offer dissenting views, and engage in open dialogue without fear of retribution. At the same time, this freedom must be accompanied by a sense of accountability. Innovation without responsibility risks turning into noise. The most successful companies establish cultures that are at once candid, collaborative, and disciplined.

And yet, cultivating such environments cannot rely solely on the instincts of charismatic leaders. Leadership, like any other professional competency, can—and should—be systematically developed. Organizations must invest in the intentional training of leaders at every level, embedding leadership development into the core of their talent strategies. Leadership is not the preserve of a few "naturals"; it is a practice that evolves through experience, reflection, and learning. Just as operational excellence requires training in tools and methods, so does transformative leadership require coaching in emotional intelligence, systems thinking, and adaptive behavior. Organizations that embrace this view build not just strong leaders but adaptive, agile cultures that can weather uncertainty and thrive amid change.

One of the most influential frameworks guiding leadership in innovation settings is transformational leadership theory. Transformational leaders are not just skilled communicators or strategists; they are deeply ethical role models who lead through inspiration and integrity. Their presence creates gravitational pull: people want to follow them, not because they must, but because they believe in the vision. These leaders are animated by purpose, and they transmit that sense of purpose through passionate storytelling, personalized coaching, and an unwavering commitment to team development. They encourage their teams to think differently, ask better questions, and stretch beyond perceived limitations. They don't simply delegate but mentor, challenge, and invest in the long-term growth of others (Goleman, 2000).

In this spirit, six leadership styles derive from core emotional intelligence competencies: coercive (demanding compliance), authoritative (mobilizing toward a vision), affiliative (prioritizing harmony), democratic (fostering participation), pacesetting (leading by example), and coaching (developing people for the future). Each style offers unique advantages in different situations. The most effective leaders, however, are not wedded to a single approach; they move fluidly among styles, responding to context and team needs. This flexibility is not a luxury; it is a necessity, particularly in innovation-driven organizations where ambiguity, volatility, and rapid change are the norm.

Ultimately, the challenge is not to choose between leadership and management, but to cultivate both. Leaders must develop the precision and operational rigor of a manager, while managers must embrace the curiosity and vision of a leader. Future-oriented leadership, then, is not a balancing act in the sense of compromise—it is a fusion. It is the ability to hold discipline and imagination in the same space—to execute today while inventing tomorrow.

Organizations that intentionally develop both sets of capabilities—managerial and leadership—position themselves for long-term relevance. In such organizations, innovation becomes not just an initiative but a mindset. Transformation becomes not just an aspiration but a discipline.

And leadership becomes not just a role but a shared responsibility embedded across the enterprise (Goleman, 2000; Kotter, 2001).

The Essential Role of Top Management in Sustaining Innovation

Innovation doesn't sustain itself. For it to become more than a scattered series of good ideas, it must be anchored by strong, consistent leadership at the very top. Without this strategic oversight from senior executives, innovation efforts often start with promises but quickly lose direction. They fizzle out—sometimes quietly, sometimes abruptly—before delivering any real transformation.

Most innovation journeys begin with a realization at the executive level: to stay relevant and competitive, the organization must embrace continuous innovation. It's a necessary condition, not a luxury. This insight often sparks energy throughout the company. People feel empowered to explore, to ideate, to push boundaries. Teams form, initiatives emerge, enthusiasm builds.

But here's the catch: innovation doesn't move on quarterly timelines. Developing real I-Cap takes time and the early stages rarely yield immediate results. For leaders used to short-term wins, this delay can test patience. If the top management's interest starts to fade, or if resources get redirected, the entire momentum stalls. Promising projects dry up, not because they failed, but because support was withdrawn too soon. That's the cost of inconsistent leadership: innovation loses its pulse.

Sustaining innovation over time requires more than momentary enthusiasm. It demands top-level ownership. Companies where innovation is treated as a peripheral or lower-level function often struggle to make real progress. Yes, innovation can emerge from the ground up, but its scale and sustainability are limited without executive sponsorship. A few motivated teams might show what's possible, but without senior backing, those sparks never turn into systemic change.

This is why long-term commitment from leadership matters. Top management must create the conditions for innovation to grow—both through strategic guidance and by embedding innovation into the organization's fabric. A dual approach works best: empowering grassroots creativity while steering it through clear strategic priorities. When done well, this creates a powerful feedback loop: bottom-up energy guided by top-down alignment.

Leadership here isn't just about vision; it's also about structure. Executives must define roles and responsibilities related to innovation. That might mean adapting existing roles, like project managers or business developers, or creating new ones such as innovation managers. Clear accountability helps connect efforts across departments, ensuring ideas don't get stuck in silos or lost in ambiguity.

Strategic oversight also means thoroughly knowing the systems. Leaders should understand the purpose and performance indicators of their innovation management systems just as deeply as they do financial metrics. They must be willing to revise those systems as needed, drawing lessons from what works and what doesn't. Integrating innovation into existing frameworks like quality or environmental management can improve cohesion and avoid duplication. Ultimately, strategic coherence across systems is essential for operational effectiveness (Boston Consulting Group, 2021).

Beyond the executive suite, the board of directors has a role to play too. While not directly involved in day-to-day innovation efforts, the board provides critical support by embedding innovation into governance priorities. Their role is to create enabling conditions: ensuring top management has the mandate, the budget, and the authority to pursue meaningful innovation. Their backing signals to the entire organization that innovation is not optional; it's central to the company's future (Barsh et al., 2008).

To sum up, top management cannot be a passive observer in the innovation journey. Their leadership must be strategic, visible, and sustained. They must set direction, structure responsibilities, maintain accountability, and communicate consistently. That's how innovation efforts evolve from hopeful beginnings to transformative, long-term capabilities with real and lasting impact.

The Chief Innovation Officer Leadership Role

In fast-moving, complex environments, strategic and structural leadership becomes a cornerstone of success, and sitting at the heart of this responsibility is the CINO. As innovation climbs the list of business priorities, the role of CINO has become increasingly central, not just as a figurehead for change, but as a catalyst guiding the organization's innovation journey.

But not every CINO is cut from the same cloth. In fact, aligning the right type of CINO to the organization's unique strategic priorities is critical.

There's no one-size-fits-all leader for innovation. Instead, six distinct leadership archetypes have emerged, each designed to meet specific innovation needs.

There's the *Researcher*, driven by scientific inquiry and long-term exploration—often focused on unlocking breakthrough discoveries. Then there's the *Engineer*, a more pragmatic figure who concentrates on making tangible improvements through incremental innovation. The *Investor* takes a more calculated approach, using financial data and risk-return logic to optimize innovation portfolios and allocate resources wisely.

On the more outward-facing side, the *Advocate* channels innovation through the voice of the customer and brand differentiation. Internally, the *Motivator* inspires creativity, fosters cultural transformation, and boosts engagement across the organization. Finally, the *Organizer* builds the operational backbone, putting systems, processes, and governance in place to ensure innovation isn't just talked about but executed effectively.

Selecting the right CINO, then, means deeply understanding both the company's strategic direction and the organizational conditions that shape how innovation happens. It's about finding a leader whose strengths align with what the business truly needs.

When it comes to what a CINO actually does, the scope is broad, but there are seven core responsibilities that consistently define the role in today's organizations.

First, they establish and support innovation best practices, standardizing methods of research and promoting openness, both internally and through external collaboration. Second, they act as capability builders, equipping employees across the company with the skills to innovate and putting systems in place to track and improve performance.

CINOs are also connectors, working across business units to accelerate high-impact projects. They scan the horizon, identifying emerging opportunities by reading market signals and detecting shifts early. At the same time, they are architects of engagement, running ideation initiatives like hackathons, crowdsourcing challenges, and other formats that unleash internal talent.

Critically, they must also manage early-stage investment. Directing seed funding to high-risk, high-reward ideas can make the difference between being disrupted or doing the disrupting. To support this, many CINOs create what are often called *innovation shelters*: protected spaces where fledgling ideas can grow outside the pressure of quarterly targets or rigid KPIs. These internal sanctuaries give innovations the breathing room they need to evolve before facing the market.

Ultimately, the CINO must walk a delicate line, championing the long-term vision while navigating the realities of day-to-day operations. It's a balancing act between agility and structure, ambition and patience. And when done right, it's one of the most transformative leadership roles an organization can have.

Strategic and Structural Leadership for Innovation

Transforming an organization is rarely a straightforward endeavor. Change initiatives often falter, not simply because of flawed strategies or insufficient resources, but because they overlook a powerful, less tangible force: human psychology. The success of any transformation depends as much on how people feel and behave as on how systems are designed or processes are executed. Too often, change efforts are built around rational plans but falter when they collide with the emotional undercurrents and cognitive biases of the people involved. This is why a deep appreciation of the psychological and behavioral dimensions of decision-making must be woven into the fabric of every transformation strategy.

Even when organizations deploy inspiring narratives, charismatic role models, incentive structures, and skill-building initiatives, change remains elusive if the human element is not fully embraced. A compelling narrative, for example, is most persuasive when it resonates on multiple levels—when it speaks to personal aspirations, offers a sense of purpose, acknowledges collective contribution, and opens the door to individual growth. When employees are invited to co-create the change story, their ownership grows, and with it, their commitment and identity within the organization grow. Simple, one-note messaging is rarely enough. It's the interplay of vision, emotion, and engagement that stirs real momentum.

Another underestimated lever is role modeling. Leaders may sincerely believe they embody the new behaviors they promote, but unless those behaviors are continually evaluated—and adjusted—they risk becoming symbolic rather than authentic. Tools like 360-degree feedback loops are crucial to ensuring leaders stay grounded and aligned with what they preach. But even the most compelling leaders can't go it alone. Change needs a fertile ground, an environment where openness to new ways of thinking and working is widespread, not just top-down.

Contrary to popular belief, money alone doesn't drive sustained behavior change. While financial incentives may generate short-term compliance, they rarely foster genuine transformation. What tends to work better? Small, unexpected rewards, and above all, a strong sense of fairness both in the process and in the outcomes. When people feel treated with dignity and transparency, they are far more likely to embrace change. Beyond incentives, capability development must include not only skills training but also psychological readiness. People need to feel emotionally equipped to adopt new behaviors. Embedding experiential learning into everyday work rather than relegating it to isolated training sessions makes new practices stick. Durable transformation stems from integrated efforts that engage hearts and minds (Aiken & Keller, 2009).

To provide structure in this complexity, a widely endorsed approach is the eight-step framework for change. It begins by igniting a sense of urgency, highlighting external pressures or threats that make the status quo untenable. This is followed by the formation of a guiding coalition, a group of influential individuals who can lead the charge. Next comes the articulation of a clear, compelling vision, one that aligns stakeholders across functions and levels. Removing obstacles is just as critical; leaders must identify and dismantle the barriers that prevent progress, whether those are structural, cultural, or behavioral. Along the way, short-term wins are essential: they energize the organization, build credibility, and maintain momentum. Ultimately, the goal is not just to implement change but to embed it, making new behaviors part of the organizational DNA so they endure beyond the initial push (Kotter, 2007).

A defining skill for today's leaders is the ability to navigate the tension between immediate performance expectations and long-term transformation goals. This balancing act requires a mindset shift from "power over" to "power with." Rather than command-and-control tactics, what's needed is collaborative leadership: one that creates a psychologically safe environment, encourages continuous learning, and nurtures meaningful participation. In doing so, leaders build morale, resilience, and commitment at every level (Taylor, 2022).

This, in turn, demands an adaptive leadership style. Gone are the days when one-size-fits-all leadership worked. Modern leaders must move fluidly between styles, attuned to context and responsive to change. They face paradoxes daily: when to assert and when to listen, when to act swiftly and when to pause for reflection. The most effective among them are those who cultivate emotional intelligence, self-awareness, cognitive empathy, and a comfort with ambiguity. These qualities are indispensable especially in times of turbulence, as seen during the COVID-19 crisis, which underscored the value of agile, emotionally attuned leadership (Jordan et al., 2022).

In short, leading transformation is not just about designing new structures or processes. It's about understanding what moves people, anticipating resistance, and adapting continuously. It's a layered endeavor that calls for both strategic discipline and human sensitivity.

To shed light on the types of challenges leaders face in complex, rapidly evolving environments, Deborah Ancona of MIT developed a powerful framework known as the *Three-Lens Model.* Particularly relevant in contexts where boundaries are blurred and the flow of information is fast and unpredictable, this model, shown in Figure 1.5, outlines three perspectives through which leaders can analyze and steer change: the strategic design lens, the political lens, and the cultural lens (Ancona et al., 2004).

The *Strategic Design Lens* sees the organization as a well-oiled machine. Here, the focus is on roles, workflows, alignment, and efficient goal setting. Leaders who favor this lens act as architects or systems engineers—concerned with clarity, coordination, and optimization. Change is driven by the need to fix inefficiencies or misalignments in how things are structured.

	Strategic Design Lens	Political Lens	Cultural Lens
Model	Organizations are machines designed (engineered) to achieve agreed-upon goals	Organizations are contexts for power and autonomy among internal stakeholders	Organizations are institutions underpinned by shared mental maps, identities, assumptions
Key Processes	Grouping (formal structure), linking, alignment, fit to the environment	Conflict, negotiation, coalition building	Meaning and interpretation, attribution, "taken for granted" (cognitive), "invested with value" (normative)
Key Concepts	Mechanical systems: goal-directed, tasks, roles, information flows, interdependence	Social systems: power, influence, networks, autonomy, interests, dominant coalition	Symbolic systems: artifacts, symbols, myths, values, assumptions, identities, subcultures
Leaders	Strategist, designer, architect; actions come through planning	Coalition builder, negotiator; actions come through power	Symbol of the culture, shaper of the culture, articulator of symbols and vision; actions come through habits
Drivers of Change	Lack of fit to the environment, internal lack of alignment	Shifts in power of stakeholders (can be influenced by changes in design, environment, or strategy)	Challenges to basic assumptions, new interpretations

Figure 1.5 The MIT Three-Lens Model. (From Carroll, J. S., The Three Lenses. Managerial Psychology, Sloan School of Management, 2006. With Permission.)

The *Political Lens*, by contrast, portrays the organization as a landscape of competing interests and power plays. Influence, coalition-building, and negotiation are the key processes here. Leaders become diplomats or brokers, navigating competing agendas and aligning divergent stakeholders. Change is often sparked by shifts in power dynamics, stakeholder priorities, or external pressures.

The third, the *Cultural Lens*, invites leaders to see the organization as a community held together by shared values, rituals, and symbolic meanings. In this view, leaders act as interpreters of meaning and custodians of purpose. Change unfolds when prevailing assumptions are challenged, when new stories and identities emerge, and when habits are reshaped over time.

To lead innovation effectively, leaders must integrate all three lenses. They need technical acumen to design coherent systems, the political savvy to manage influence and navigate resistance, and the cultural sensitivity to reframe beliefs and inspire collective movement. Innovation leadership, therefore, isn't just about having a great idea or building the right team. It's about crafting the conditions—structural, political, and cultural—under which new ideas can take root, grow, and flourish (Ancona et al., 2004; Di Fiore, 2014; Lovric & Schneider, 2019).

References

Aiken, C., & Keller, S. (2009). The Irrational Side of Change Management. *The McKinsey Quarterly.*

Ancona, D., Kochan, T., Scully, M., Van Maanen, J., & Westney, D. E. (2004). Managing for the Future: Organizational Behavior and Processes (3rd ed.). *South-Western College Publishing.*

Barsh, J., Capozzi, M. M., & Davidson, J. (2008). Leadership and Innovation. *The McKinsey Quarterly.*

Boston Consulting Group (2021). The CEO Innovation Agenda.

Corbett, A. (2018). The Myth of the Intrapreneur. *Harvard Business Review.*

Di Fiore, A. (2014). A Chief Innovation Officer's Actual Responsibilities. *Harvard Business Review.*

Goleman, D. (2000). Leadership That Gets Results. *Harvard Business Review.*

Govindarajan, V., & Desai, J. (2013). Recognize Intrapreneurs Before They Leave. *Harvard Business Review.*

Hamel, G., & Zanini, M. (2014). Build a Change Platform, Not a Change Program. *McKinsey & Company.*

ISO (2020). ISO 56000:2020 – Innovation management – Fundamentals and vocabulary. *International Organization for Standardization.*

ISO (2024). ISO 56001:2024 – Innovation management system – Requirements. *International Organization for Standardization.*

Jordan, J., Wade, M., & Yokoi, T. (2022). Finding the Right Balance - and Flexibility - in Your Leadership Style. *Harvard Business Review.*

Kanter, R. M. (2006). Innovation: The Classic Traps. *Harvard Business Review, 84*(11), 72.

Kotter, J. P. (2001). What Leaders Really Do. *Harvard Business Review.*

Kotter, J. P. (2007). Leading Change: Why Transformation Efforts Fail. *Harvard Business Review.*

Lovric, D., & Schneider, G. (2019). What Kind of Chief Innovation Officer Does Your Company Need? *Harvard Business Review.*

Schumpeter, J. A. (1911). The Theory of Economic Development. *Harvard University Press.*

Schumpeter, J. A. (1942). Capitalism, Socialism and Democracy. *Harper & Brothers.*

Taylor, B. (2022). How Leaders Can Balance the Needs to Perform and to Transform. *Harvard Business Review.*

Zenger, J., & Folkman, J. (2014). Research: 10 Traits of Innovative Leaders. *Harvard Business Review.*

Chapter 2

Innovation Strategy: A Framework for Sustainable Competitive Advantage

Strategy According to ISO 56000 Series Requirements

In any organization that seeks to innovate deliberately and sustainably, strategic direction must be more than a statement of intent; it must be a coordinated effort, shaped by ambition, guided by clarity, and sustained by the intelligent use of resources. The International Organization for Standardization (*ISO*) *56000* series frames this direction with precision, outlining how innovation should be rooted in a shared sense of purpose, with clear objectives and sufficient human and material resources to bring those objectives to life.

According to *ISO 56000* (ISO, 2020), and specifically *Clause 4.1.3*, a well-articulated innovation strategy does more than just align internal initiatives. It allows for the systematic allocation of roles, the distribution of resources, and the establishment of governance mechanisms to ensure that innovation activities are prioritized thoughtfully and evaluated rigorously. A clearly defined strategic direction also serves as a powerful communication tool. It helps internal stakeholders understand the organization's ambitions and expectations, while signaling to external actors that innovation is not peripheral but central to the organization's identity and growth trajectory.

The strategic direction described by *ISO 56000* is then further elaborated in *ISO 56001* (ISO, 2024), which provides more detailed operational

DOI: 10.4324/9781003743583-2

guidance for translating intent into structured planning. *Clauses 6.1.1 and 6.1.2* lay out how organizations should address the dual challenge of seizing opportunities and mitigating risks within their innovation systems. Rather than treating risk as something to avoid entirely, ISO encourages organizations to define their risk appetite, accept a certain level of uncertainty, and make deliberate choices about which risks are worth taking in the pursuit of innovation. These choices, however, must not remain abstract. They must be embedded into the operational processes of the innovation management system and revisited periodically to assess relevance and effectiveness. Strategic planning, therefore, is not a one-off effort but an ongoing exercise in learning, adapting, and refining.

Central to this process is the establishment of measurable innovation objectives. These goals should flow naturally from the broader innovation policy and vision, ensuring consistency throughout the organization. They must also be realistic, clearly communicated, and widely understood by those responsible for executing them. Every objective should come with a corresponding set of actions, timelines, required resources, and criteria for evaluation. This clarity transforms strategic aspirations into a living operational reality, where all contributors understand not only what they're working toward but also how success will be recognized and measured.

ISO 56001 also brings attention to the importance of designing organizational structures that can support both creativity and disciplined execution. *Clause 6.3* emphasizes that organizations should not rely on rigid hierarchies or legacy systems when building innovation capabilities. Instead, they must remain open to rethinking how structures enable or constrain innovation. This might mean establishing new units focused on disruptive technologies or embedding innovation practices within existing teams through cross-functional collaboration. In either case, top management is responsible for ensuring that the organization is structurally equipped to translate innovative ideas into results.

Portfolio management, addressed in *Clause 6.4*, introduces an additional strategic layer (see Figure 1.3 "A Small-I to Big-I portfolio approach."). Here, innovation is treated as a dynamic set of initiatives that must be continuously assessed, adjusted, and aligned with evolving business goals. Rather than letting innovation efforts drift, organizations are encouraged to actively manage their portfolios, weighing the potential value and risk of each initiative, allocating resources accordingly, and making hard decisions about what to continue, redirect, or stop. This kind of disciplined flexibility ensures that innovation remains responsive to real-world conditions without losing sight of its long-term aims.

Taken together, the *ISO 56000* and *ISO 56001* standards offer a comprehensive strategic compass for innovation management. They do not prescribe a single path forward but instead provide the principles and practices necessary to build, maintain, and evolve an innovation strategy that is both visionary and executable. When properly applied, these standards help organizations embed innovation into their very structure not as a separate department or occasional project but as an ongoing, strategic capability that is continuously monitored, refined, and aligned with a changing business environment.

In the end, strategy in innovation is not only about vision or ambition. It is about designing systems that can hold that ambition, carry it forward, and translate it into results. *ISO 56000* offers a roadmap for doing just that: building alignment, structuring decision-making, managing uncertainty, and enabling the kind of adaptability that is essential for long-term success in dynamic and competitive markets.

Strategic Context for Innovation and Technological Progress Over the Last Four Decades

In recent times, technological progress has evolved in terms of developing key technologies through innovative approaches such as closed innovation, open innovation, and innovation ecosystems. These developments have led to a period of great change in the dynamics of innovation.

As a result, business innovation has become a strategic aspect to compete in global markets. The concept of innovation and entrepreneurship has undergone several definitional changes over the past four decades, and there have been six major waves of innovation that have led to the current approach to innovation.

The first was the beginning of the global information age in the early 1980s, an era that introduced new industries and threatened old ones. Computing was starting to move from mainframe computers to consumer and desktop products. Companies such as Apple Computer set up garages in Silicon Valley, which was the new base for product innovation in the United States. IBM took inspiration from Apple's model by developing its PC in Boca Raton, Florida, and because it was free from many corporate constraints, it was able to outperform Steve Jobs' startups in the early stages. High-quality Japanese products, such as the Sony Walkman and Toyota cars, reflected not only good design but also innovations in manufacturing processes that forced the American giants to create programs to generate new ideas more quickly. "Total Quality Management" became a benchmark for competition.

The second wave began due to the pressure of restructuring during the takeover scare of the late 1980s. Buyout groups were competing against traditional companies to try to unlock the value of underutilized assets. "Shareholder value" became a constant requirement. In Europe, restructuring was associated with the privatization of state-owned enterprises that became exposed to pressure from capital markets. Software was emerging as a driving force for innovation and the strategic value of IT was being extolled. For example, American Airlines' Sabre reservation system shows successful process innovation as a separate business from the parent company. Companies created startup divisions to ensure they were leveraging the value of their own ideas rather than allowing a new venture like Microsoft to emerge outside the company's boundaries. Financial innovations became critical, and takeovers through derivatives and other forms of financial engineering were leveraged. The restructuring era also favored products that could already be global: after rejecting a takeover bid in the late 1980s, Gillette launched its Sensor Excel shaving technology in the early 1990s. It was promoted in the same way around the world with a single advertising message.

The third was the digital craze of the 1990s. The creation of the World Wide Web pushed many companies to seek radical new business models. Traditional companies were in danger of going bankrupt and becoming extinct. The focus was on capital markets and not so much on customers. Companies proved valuable even without profits or revenues. An example of this dynamic occurred at the end of this third wave: in 2001, American OnLine (AOL) acquired Time Warner and became AOL Time Warner. Because of its poor skills in the mass media market, efforts were made to convert its stock price into tangible assets, but AOL ended up destroying value rather than creating innovation.

In the late 2000s, AOL and Time Warner announced their merger, which was valued at $350 billion. There were high expectations about how the two companies would complement each other. At the time, AOL had a dominant position in the dial-up Internet industry, while Time Warner owned the content but had limited knowledge of the Internet. However, shortly after the merger, there were discrepancies between the two companies due to their different corporate cultures—AOL had a more sales-oriented culture, while Time Warner had a more traditional corporate culture. This led

Time Warner to reject AOL's proposals and look for alternative partners to distribute their online content. With the spread of broadband, AOL was unable to count on Time Warner as a partner and began to develop its own cable business; this ended with the merger failing, and the combined value of both companies was less than before the merger. Subsequently, in 2009, Time Warner created Time Warner Cable, and in December 2009, AOL and Time Warner split completely. AOL was then acquired by Verizon in 2015, and Time Warner by AT&T in 2016. To avoid bankruptcy, AOL and Time Warner could have made compromises to respect and integrate their respective corporate cultures, or they could have chosen a dominant culture to stick to. Unfortunately, neither of these were implemented, and this created dysfunction in the merger.

The fourth wave began in the early 2000s. It took on a different tone after the dot-com crash and cost-cutting measures due to the global recession. After becoming wary of technological hype, companies refocused on organic growth. Surviving companies such as General Electric and IBM began to integrate innovation as a corporate theme. In this wave, there was a major focus on new products designed to offer users new features and functionality to meet emerging needs. Customers and consumer markets returned to the center of corporate interests; they had been temporarily sidelined by other interests. Companies sought new market niches to grow their businesses and did not try to create new, big-looking initiatives, introducing them in completely different business contexts. Innovations of this era included Apple's iPod and Procter & Gamble's Swiffer. The former was an innovation that focused on design and ease of use; the focus was not only on technological development. The second was an innovation that developed through a customer-centric approach.

The fifth wave of innovation developed in the second decade of the new century; it was born from the convergence of software and hardware. The growth in software quality combined with the availability of cheap and powerful processors and sensors, that is more reliable electronic components allowed the creation of cyber-physical systems (also known as Industry 4.0 first and Industry 5.0 later). For example, General Electric's priority was the production of jet engines, trains, and turbines that use software and sensors to measure and optimize their performance. The vision was that the union of software and hardware would lead to major technological changes.

Tesla Motors' Model S sedan is an example of this convergence—it's not just that the Model S is a smooth, quiet ride; it's also that it can be seen as a gadget. Its wireless connectivity supports streaming Internet radio stations and allows the driver to unlock the car using their smartphone. If it breaks down, owners don't have to go to a mechanic to have it fixed. Instead, the company says, "While you sleep at night, a software engineer will connect to the car, find the problem, and fix it." This changes everything in the world of transportation—the Model S is an example of all the new software-based devices that could create a huge opportunity. These devices give companies the ability to redesign their consumer products, just as Tesla redesigned the car and Apple redesigned the phone. In the latter case, the most radical model of change is the iPhone; the hardware-software combination has made the cell phone a pocket computer and shifted the balance of power in communications technology from Asia and Europe to the United States. Another example is Google, which has improved its position in America with its Android software, phones, and tablets. Apple has become the world's most valuable company, and Google's dominance in software is the envy of the industry.

Several technological streams have affected both corporate proprietary technologies and business models, such as the artificial intelligence (AI) generating Google's self-driving capabilities; Google Allo, a smart messaging app; and IBM Ross, the first artificially intelligent attorney. Other examples include 3D printers generating bioprints of tissues and organs and the Internet of Things (IoT) generating the Amazon Web Services' IoT Button, Amazon's programmable Dash Button that can be integrated into any product or good as a customizable new version of a one-touch reordering gadget. Not to mention the enabling IoT influence toward the already widespread "sharing economy" or the impact of AI and robotics on the social media market. Web2 also gave rise to web apps, self-publishing platforms like WordPress, as well as social media sites. Examples of Web2 sites include Wikipedia, Facebook, Twitter, and various blogs, which all have transformed the way the same information is shared and delivered.

The current sixth wave of technological innovation is a collection of cutting-edge technologies that are set to impact multiple industries. One such technology is Web3, a term often used to describe an emerging iteration of the World Wide Web built on blockchain based infrastructures and decentralized protocols. In principle, Web3 combines smart contracts, token based incentives, and distributed governance models to reduce reliance on centralized intermediaries and to enable new forms of digital ownership and coordination. In practice, however, the extent to which it truly decentralizes control depends on design choices, regulatory constraints, and the maturity of supporting ecosystems such as identity, custody, security, and user experience. For this reason, many current implementations should be read as early experiments that shift certain functions away from dominant platforms, rather than as a definitive replacement of today's web. Companies such as the Dutch blockchain LTO Network have begun exploring these possibilities through initiatives sometimes described as NFT 2.0, aimed at improving the management, traceability, and exchange of non fungible tokens in more operationally robust ways.

Another newly developed technology is artificial intelligence (AI) and machine learning, which have emerged not merely as tools of efficiency, but as foundational forces reshaping the way industries operate, decisions are made, and value is delivered. No longer confined to advanced research, these technologies are now embedded in sectors as diverse as healthcare, logistics, finance, transportation, and manufacturing.

Finally, a rapidly evolving frontier within AI is agentic AI, namely systems designed to pursue goals, plan multistep actions, and execute tasks with limited supervision. Rather than stopping at prediction or content generation, agentic AI extends AI into coordinated action across tools, workflows, and organizational processes, potentially transforming software from systems of record into systems of execution, while raising new requirements for governance, accountability, and control.

What's particularly striking is how this transformation transcends technical domains and penetrates the very structure of capitalism itself. AI is not simply changing how things are made; it's redefining how value is created, customized, and consumed. We are witnessing a profound shift—from the rigid, industrial-era logic of Fordism, built around mass production and standardization, to a new digital paradigm grounded in mass customization and responsive intelligence.

Whereas the Fordist model excelled in producing identical products at scale, today's AI-powered systems allow companies to deliver personalized experiences at industrial efficiency. Algorithms trained on vast streams of customer data can now interpret preferences, predict needs, and tailor offerings—sometimes in real time. This means that the dream of creating a unique solution for every customer is no longer an aspirational fantasy but an operational reality. The tension between customization and scalability, once a defining trade-off in economics, is being dissolved by AI.

In this new paradigm, *mass customization* becomes the hallmark of competitiveness. Businesses can generate thousands—or millions—of variations of a product or service, each one fine-tuned to a specific individual, all while keeping costs contained. This fusion of personalization and efficiency enables companies to target pain points with remarkable precision, creating a level of customer value that was inconceivable in the age of one-size-fits-all solutions.

Equally important is the digital paradigm's capacity for continuous learning. Companies are no longer limited to periodic market research or after-the-fact analytics. With AI integrated into their operations, they can collect and process real-time customer data, uncovering insights that are immediately actionable. This transforms how businesses relate to their customers from distant observers to responsive partners. Preferences are no longer guessed; they are observed, learned, and adapted to, instantly. In this context, data becomes not just a resource but a living feedback loop, fueling an ever-evolving product and service design process.

In sum, the digital paradigm powered by AI is not just an upgrade to older production systems; it's a foundational shift in how businesses operate and compete. It offers a model of production and consumption that is deeply attuned to the individual, yet operationally streamlined. It challenges the legacy of mass standardization and replaces it with adaptive intelligence, embedded into every layer of value creation.

Yet even with such transformative potential, the journey from innovation to sustainable advantage remains uneven. In today's hypercompetitive global markets, the ability to innovate consistently and meaningfully—across products, processes, and business models—is critical. But many companies still struggle to translate their innovation investments into systemic internal change. Even firms renowned for their inventive spirit—Apple, 3M, Google, Microsoft, Meta, Amazon, and Tesla—face ongoing difficulties in embedding innovation deep into their organizational DNA. It is not a matter of resources but of integration: innovation must not be a separate initiative but a structural capability woven into everyday practice.

To reach a tipping point, organizations must move beyond isolated research and development (R&D) departments or occasional breakthroughs. They must cultivate innovation density—a rich and interconnected set of capabilities, incentives, and cultural norms that promote sustained creativity and agile execution. Achieving this requires a broader strategic lens—one that recognizes the company not as an isolated actor but as part of a larger innovation ecosystem.

This ecosystem includes a complex web of stakeholders: entrepreneurs, venture capitalists, researchers, policymakers, corporations, and civil society (see section "Innovation Ecosystems and the Collaboration for Knowledge Integration" in Chapter 6). When these players interact meaningfully—sharing knowledge, exchanging talent, and pooling resources—they generate the kind of "new-to-the-world" ideas that fuel long-term innovation leadership. The most resilient and forward-thinking companies are those that actively engage with these ecosystems, positioning themselves as contributors, collaborators, and connectors in an ever-expanding innovation landscape.

But such engagement is not a one-time alignment. It is a dynamic, ongoing process. Ecosystems evolve. Markets shift. Technologies mature. And so must the strategies and roles of every stakeholder involved. To remain relevant, every player—whether a startup founder, a corporate R&D leader, or a public policy architect—must be willing to adapt, rethink, and reimagine their place in the innovation chain.

In this context, innovation is no longer just an outcome. It becomes a mode of being—a strategic posture, a shared commitment, and a continuous conversation across boundaries and disciplines. Those who embrace this shift won't just survive the next wave of disruption; they'll shape it.

Cultural Foundations of Innovation Strategy

In the era of AI where technology reshapes not just industries but the very nature of value creation, innovation has become a non-negotiable imperative for companies seeking relevance, resilience, and long-term success. No longer a luxury or a niche capability, innovation today defines whether organizations survive or disappear. This explains why so many companies pour significant resources into improving both the speed and the quality of their innovation outcomes, yet success is not guaranteed by investment alone. The cultural foundations supporting these innovation efforts often determine whether they thrive or stagnate.

At its core, innovation is best understood not merely as an invention, nor as the exclusive domain of advanced technology. Rather, it is a journey: a process that takes ideas from initial conception through to meaningful impact. And impact, in this sense, extends beyond financial return. It encompasses societal and environmental contributions as well; proof that innovation is ultimately measured not just in revenues but in relevance and responsibility.

To drive transformative innovation, organizations must go beyond refining the familiar. A culture committed to originality, not just efficiency, is essential (see section "Establishing a Culture for Innovation" in Chapter 6). Modern entrepreneurial thinking prioritizes the pursuit of the radically new: creating products, services, or models so distinct that alternatives either do not exist or fall short in comparison. This vision of innovation challenges the traditional fear of monopolies as inherently dangerous entities that stifle competition and harm consumers. Instead, it reframes monopoly status—when born of innovation rather than coercion—as a desirable outcome, particularly for startups and young ventures aiming to redefine entire sectors.

These "good monopolies" are not the result of protectionist practices or unfair market manipulation. They arise from unique, often daring innovations that unlock new markets or revolutionize existing ones. Their strength lies in building value propositions that others cannot imitate. Companies that achieve this level of uniqueness often benefit from network effects, where the utility of a product increases with each new user, and from economies of scale, which allow them to reduce costs while maintaining high quality and performance. This virtuous cycle reinforces their market position while enabling further reinvestment in innovation.

In this light, monopolistic leadership becomes not a threat to the economy but a catalyst for growth. Freed from the pressure of short-term survival, these organizations can think in decades, not quarters. They channel their resources into advanced research, pioneering solutions, and long-term bets on what their industry—and society—might need next. In doing so, they help define the next generation of markets rather than merely participating in them. They solve problems others haven't yet dared to tackle and address needs that were previously unmet or even unimagined.

Yet this model is not without critics. Detractors argue that monopolies, regardless of origin, can become complacent and detrimental to consumers, reducing choice and inflating prices. But this critique applies more readily to the monopolies of the past—those that maintained dominance by suppressing competition, not through innovation, but through legal and economic strong-arming. By contrast, today's dynamic monopolies emerge from creative destruction. They rise not by closing doors but by opening entirely new ones. They do not merely outperform the competition; they make previous models obsolete (Thiel & Masters, 2014).

This strategic cultural shift also calls for a broader understanding of technological progress. Organizations that truly want to lead innovation must train their gaze not only on individual products or quarterly goals but also on the underlying forces reshaping entire industries. They must develop the capability to interpret these shifts to map the architecture of change and position themselves accordingly.

This is especially urgent in the current phase of global transformation. The digital paradigm now eclipsing the old industrial logic represents a fundamental break from the Fordist approach of mass production and standardization. In its place, we are witnessing the rise of hyper-personalization, digital intelligence, and flexible systems that can tailor value to the individual while preserving scale. The technologies powering this shift—from advanced automation and AI to real-time data analytics and IoT—are not just tools. They are new building blocks for how economies operate and how companies engage with customers.

This transition demands more than technical adoption. It calls for a reimagining of culture itself: a culture that rewards exploration over repetition, dialogue over hierarchy, and long-term vision over short-term gain. In such a culture, innovation is not a side initiative but an ethos woven into decisions, relationships, and ambitions across the organization.

The cultural foundation of innovation, then, is about far more than values printed on a wall. It is about building an environment where radical ideas are not only welcomed but also structured and resourced, where the fear of failure does not outweigh the desire to try, and where innovation is measured not simply by profit but by the broader impact it creates. In this context, innovation becomes a societal force, not just a business strategy.

Innovation Strategy Framework

Strategic Vision and Alignment in Innovation Management

In today's fast-changing landscape, organizations are often driven by bold aspirations—be it delighting users through exceptional experiences, accelerating growth at scale, embracing digital transformation, delivering on sustainability commitments, or even reshaping society in meaningful ways. But such ambitions rarely find fulfillment through business-as-usual. Traditional plans and established capabilities, no matter how refined, are often ill-equipped to handle the ambiguity and complexity that define these higher goals.

This is where innovation becomes more than a buzzword; it becomes a strategic imperative. Especially in times of heightened uncertainty, the ability to innovate with purpose and direction is what separates companies that merely adapt from those that lead. But innovation doesn't flourish in a vacuum. It needs structure. It needs intention. And most of all, it needs to be tied—consciously and explicitly—to the organization's broader strategic vision. Even the most promising ideas can wither without the support of a well-articulated strategic foundation (Pisano, 2015).

An innovation strategy, when thoughtfully developed, becomes much more than a roadmap; it becomes a source of clarity for the entire organization. It helps business units determine where to focus, what types of opportunities are worth pursuing, and how best to translate vision into action. At its core, it defines how innovation will serve as a bridge between long-term ambition and day-to-day execution. It provides a shared language and a common sense of purpose, ensuring that innovation is not just encouraged but understood and championed across the board.

Creating such a strategy begins with seeing clearly, both outward and inward. Externally, companies must stay attuned to shifting behaviors, technological breakthroughs, emerging models of value creation, policy disruptions, and cultural transformations. Internally, they must honestly assess their readiness to innovate: their capabilities, resources, leadership mindset, and organizational norms. Without this dual awareness—outside-in and inside-out—even the best strategy will fall short.

To make innovation actionable and aligned, four strategic dimensions come into play. First, *envisioning value* means painting a vivid picture of what the future could look like for users, customers, communities, employees, and shareholders alike. These visions are not abstract ideals but grounded aspirations that guide short- and long-term priorities.

Second, *framing opportunities* is about mapping the terrain, clarifying which areas deserve exploration, how ambitious the organization should be, and where its efforts are most likely to generate traction. It's about balancing realism with boldness and ensuring that limited resources are funneled toward ideas with real promise.

Third, *defining approaches* entails making strategic choices: should innovation be driven by emerging technologies or unmet customer needs? Should efforts be collaborative or run independently? Here, organizations must articulate not just what they will do but how they will do it and commit real resources to back those choices.

Finally, *ensuring governance* provides the mechanisms for sustained momentum. This includes defining decision-making criteria (e.g., value potential, feasibility, environmental and social impact, investment needs), tracking performance, and embedding innovation KPIs into the broader enterprise scorecard. Governance is not about control; it's about direction, accountability, and learning in motion (Reeves et al., 2012; Pisano, 2015).

To better illustrate how innovation strategies operate across different contexts and degrees of uncertainty, Figures 2.1 and 2.2 offer powerful conceptual maps of the solution-problem space.

To truly harness innovation, it's essential to understand where innovation happens, not just in terms of industry or technology, but in the space between problems and solutions. Figure 2.1 maps this terrain and highlights three archetypal innovation paths that organizations commonly follow: technology-push, market-pull, and design-driven approaches. Each path reflects a distinct interplay between the novelty of the problem being addressed and the originality of the solution being proposed.

Let's begin with the technology-push model. Here, innovation originates from scientific exploration or technical advancement, often before a clear need has been expressed by the market. These innovations push the boundaries of what's possible, bringing forth radical outcomes that can eventually find fertile ground, even if the problems they solve are not yet fully defined. This approach is common in research-driven fields, where breakthroughs such as advanced materials, biotechnology,

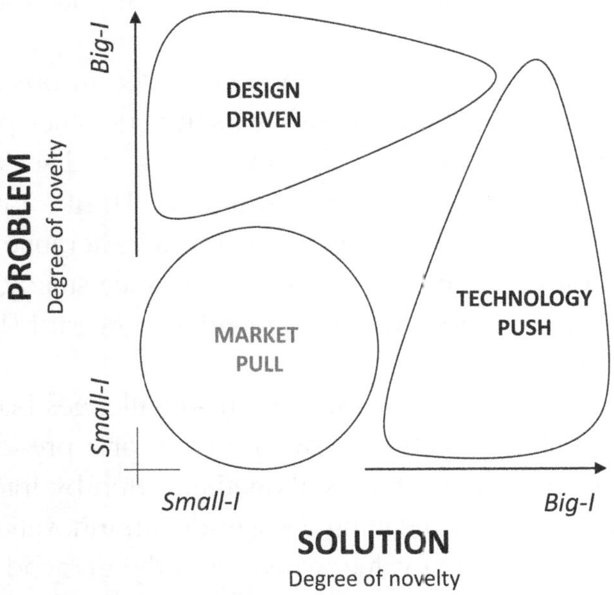

Figure 2.1 Innovation in the solution-problem space.

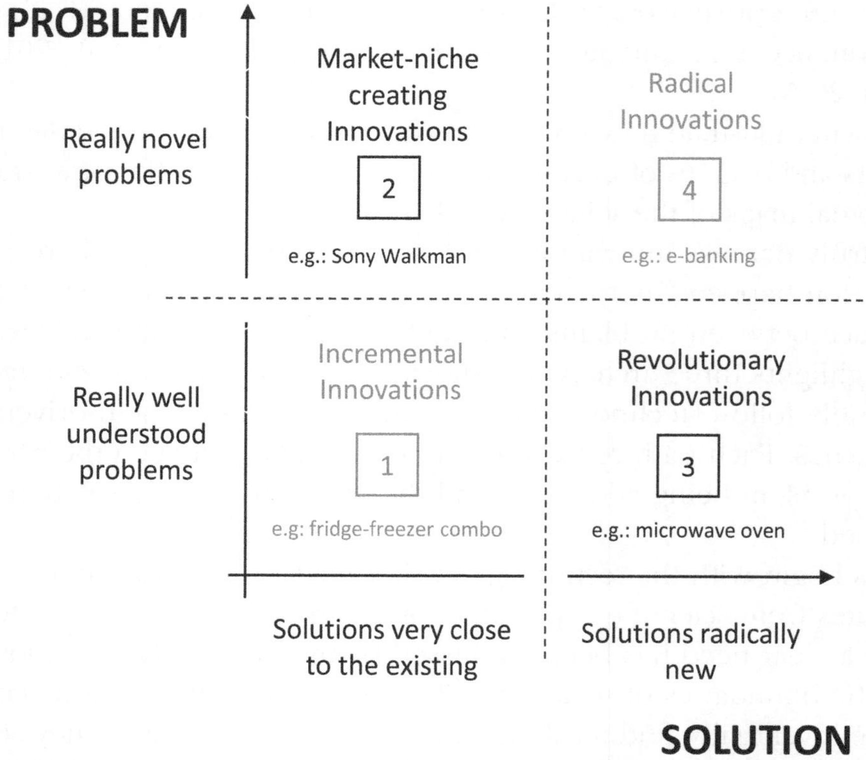

Figure 2.2 Solution–Problem Innovation Matrix.

or AI emerge from labs before their commercial applications are fully understood.

In contrast, the market-pull approach is grounded in observation and responsiveness. It begins by recognizing existing customer problems—needs that are already visible and often well-understood—and seeks to meet them through increasingly refined or efficient solutions. Think of it as a continual dialogue with the user, where innovation evolves in iterative steps, guided by feedback and proven demand. It tends to produce steady, incremental advances, but when executed well, it can lead to powerful shifts in customer experience and value delivery.

The third path—design-driven innovation—challenges both ends of the spectrum. It doesn't begin with a known problem or a pre-existing solution. Instead, it seeks to reframe problems altogether, often by interpreting subtle shifts in behavior, culture, or meaning. Design-driven innovation is about creating entirely new spaces of value, where both the demand and the offering are shaped in parallel. It often relies on visionary thinking, deep empathy, and the blending of technology, business models, and cultural insight.

These innovations don't just solve problems; they create new desires and reshape expectations.

Together, these three approaches show that innovation isn't a one-size-fits-all activity. It lives in the tension between what is technically possible, what people need, and what hasn't yet been imagined. A mature innovation strategy knows when to push forward with invention, when to listen and respond, and when to reimagine the very context in which people live, work, and consume.

If Figure 2.1 maps the broad terrain of innovation activity, Figure 2.2 offers a more structured lens for understanding the types of innovation outcomes that can emerge based on the novelty of both the problem and the solution. This Solution-Problem Innovation Matrix breaks the landscape into four distinct quadrants, each revealing a different kind of strategic opportunity and challenge.

In Quadrant 1, we find incremental innovations. These are evolutionary rather than revolutionary. They solve familiar problems with familiar tools, enhancing what already works. Consider the classic example of the fridge-freezer combo, a modest yet meaningful improvement over separate appliances. It's the domain of continuous improvement, efficiency gains, and minor user experience upgrades. Often overlooked, incremental innovations are the lifeblood of sustained competitiveness in mature industries.

Quadrant 2 represents market-niche creating innovations. These arise when companies take an existing or emerging solution and apply it to a newly recognized problem. A famous example is the Sony Walkman. Technologically, it wasn't groundbreaking; Sony simply miniaturized and reconfigured components that already existed. But in doing so, they tapped into a latent desire: personal, mobile music consumption. At the time, people listened to music at home, often in shared spaces. The Walkman created a new niche by allowing users to take music with them, thereby transforming not just consumption, but identity and lifestyle.

Moving to Quadrant 3, we encounter revolutionary innovations, radically new solutions applied to well-known, persistent problems. One such innovation is the microwave oven, which addressed the universal need for faster cooking in an entirely new way. These kinds of breakthroughs tend to gain traction faster because the problem is already recognized and urgent; what's new is the approach to solving it.

Finally, Quadrant 4 is home to radical innovations, those rare but transformative moments where both the problem and the solution are new.

These innovations tend to emerge from deep technological shifts and new patterns of behavior. A classic example is e-banking, which redefined not just how financial services were delivered but what banking meant in a digital world. These innovations often face significant resistance, as they require changes in regulation, user trust, and ecosystem structures. Yet, when successful, they redefine entire industries.

One key insight from this matrix is that as we move from incremental to radical, the level of uncertainty—and potential value—increases dramatically. So does the challenge of adoption. A robust innovation strategy doesn't simply chase the upper-right quadrant. Instead, it balances the portfolio, leveraging the reliability of incremental innovation while also allocating resources to high-risk, high-reward bets that can future-proof the organization (see Figure 1.3 "A Small-I to Big-I portfolio approach.").

Business Model Innovation and Value Creation

In fast-moving markets shaped by constant disruption, having a deep, nuanced understanding of the business model is no longer optional; it's fundamental to achieving sustained advantage. A business model isn't just a blueprint for making money; it's the connective tissue that links an organization's capabilities to customer needs, competitive positioning, and long-term value creation. Foundational thinkers like Michael Porter and Peter Drucker offer timeless perspectives that continue to shape how we examine the mechanics—and the strategy—behind value.

Porter's work, especially his value chain model (1985), breaks down a company's activities into two main sets: primary functions and support functions. Primary activities include visible, customer-facing operations like inbound logistics, manufacturing, marketing, and after-sales service. Support functions, on the other hand—such as technology development, HR, and organizational infrastructure—work in the background but are equally essential. When these components operate in harmony, inputs (such as resources and capabilities) are seamlessly transformed into outputs that deliver value to the customer. This process is what ultimately fuels competitive advantage.

Meanwhile, Drucker (2012 [1954]) offers a more holistic lens through his five core elements of a business model: the customer, the value proposition, the product or service offering, the value creation mechanism, and the value appropriation mechanism. These elements encourage leaders to see beyond the internal configuration of the firm and to focus on how the business aligns with external realities. When we connect Drucker's framework to Porter's model, the pieces start to form a coherent system. The customer becomes the end beneficiary of all primary value-chain activities. The value proposition and the offering represent the outcomes of a well-functioning system, while the value creation mechanism reflects the internal processes that convert resources into benefits. Finally, the value appropriation mechanism addresses how the company captures a share of that value, often through its ability to leverage support functions efficiently (see section "Business Model Innovation as an Adaptive Strategy" in Chapter 4).

In essence, while Porter grounds us in the operational anatomy of value creation, Drucker challenges us to align that anatomy with strategic intent and market fit. Together, they offer a dual perspective: one inward, emphasizing structure and execution; the other outward, focused on relevance, purpose, and adaptability.

This alignment becomes even more crucial in the context of digital transformation. In an age where uncertainty is the rule rather than the exception, traditional long-term planning approaches often fall short. Large-scale transformations are notoriously risky-slow, expensive, and frequently misaligned with evolving customer behavior. To counter this, organizations are turning to discovery-driven planning (DDP): a flexible, iterative, and experimental mindset that prioritizes learning over rigid execution.

Rather than committing upfront to fully defined roadmaps, companies adopting this method focus on leveraging existing strengths—such as deep customer knowledge, operational capabilities, and internal talent pools—to test, learn, and adapt in real time. This approach minimizes disruption while enabling the evolution of new digital business models in a controlled yet responsive way.

At the heart of this method lies the framework of DDP, developed specifically for innovation initiatives marked by high uncertainty. Instead of beginning with revenue projections and resource plans, DDP encourages teams to start by defining what success actually looks like, often through tools like the reverse income statement, which builds financial assumptions from the bottom up. The method also emphasizes benchmarking feasibility, mapping operational requirements, and identifying critical checkpoints for reflection and recalibration. Influential in shaping the Lean Startup movement (see section "Managing Disruptive Technologies: Contrasting Approaches of Large Enterprises and Innovation-Driven Enterprises (IDEs)" in Chapter 5), DDP reinforces the importance of experimentation, real-time learning, and adaptive planning as essential tools for navigating ambiguity (Gallo, 2017; McGrath & McManus, 2020).

Stepping back to look at business models from a broader perspective, four essential dimensions shape how innovation unfolds within them. These include the *nature of what is offered*—namely, the products or services the organization chooses to provide; *the timing of key decisions*, which can influence flexibility and responsiveness; *the assignment of decision-making authority* or who within the organization is empowered to make those calls; and, finally, *the underlying rationale behind those choices*—why certain paths are pursued over others, based on strategic intent or emerging evidence. Each of these dimensions reflects a distinct lever of design, and together they form the backbone of any effort to reimagine how value is created and delivered in uncertain environments (Girotra & Netessine, 2014).

Each of these dimensions represents a design lever. For instance, a company might decide to focus exclusively on one core offering or pursue diversification. It might delay certain strategic decisions until more data becomes available or empower specific teams to make rapid calls. These are not just operational choices; they are strategic reconfigurations that determine how flexible, scalable, and competitive a business can be in uncertain conditions.

True innovation in business models, therefore, is not simply about launching something "new." It's about rethinking the logic of value creation itself: who benefits, how, and under what conditions. It means challenging existing assumptions, revisiting the architecture of decision-making, and redesigning internal systems to support agility and responsiveness.

In the end, the organizations that will thrive are those that design business models not just for efficiency but for resilience. Models that don't just cut costs but anticipate change, embrace experimentation, and remain attuned to evolving customer expectations. Discovery-driven approaches don't eliminate risk, but they manage it more intelligently by replacing rigidity with reflection, and assumption with evidence. In doing so, they empower organizations to evolve alongside the very markets they aim to shape.

Market Dynamics and Strategic Trade-Offs

Disruptive Change and First-Mover Advantage

In the ever-evolving theater of modern markets, companies are not merely participants—they are agents of transformation, constantly navigating a shifting terrain shaped by innovation, competition, and institutional influence.

Strategic advantage no longer stems only from size or scale but from a company's ability to recognize emerging opportunities, move decisively, and balance competing forces such as innovation incentives, market dominance, and the costs of disruption.

A central insight from Schumpeterian economics is that innovation carries with it an inherent trade-off. In his vision, progress is inseparable from creative destruction—a cycle in which novel solutions overtake the old. Yet for firms to engage in high-risk, long-term innovation, they must be granted temporary monopolies, a window of opportunity to reap the rewards of their investments before imitators arrive. As illustrated in Figure 2.3, this delicate balance unfolds across two possible paths. One allows for sustained monopoly rents, encouraging bold investment in R&D and enabling breakthrough innovations to emerge and mature. The other offers only a brief window of profitability, resulting in swift imitation and reduced incentives for innovation. While the former stimulates the innovation engine, it can delay the return to competitive market conditions, incurring societal costs in the form of reduced consumer choice or inflated prices. Conversely, the latter fosters competition but may discourage radical innovation, as firms may perceive the payoff as too uncertain or too short-lived.

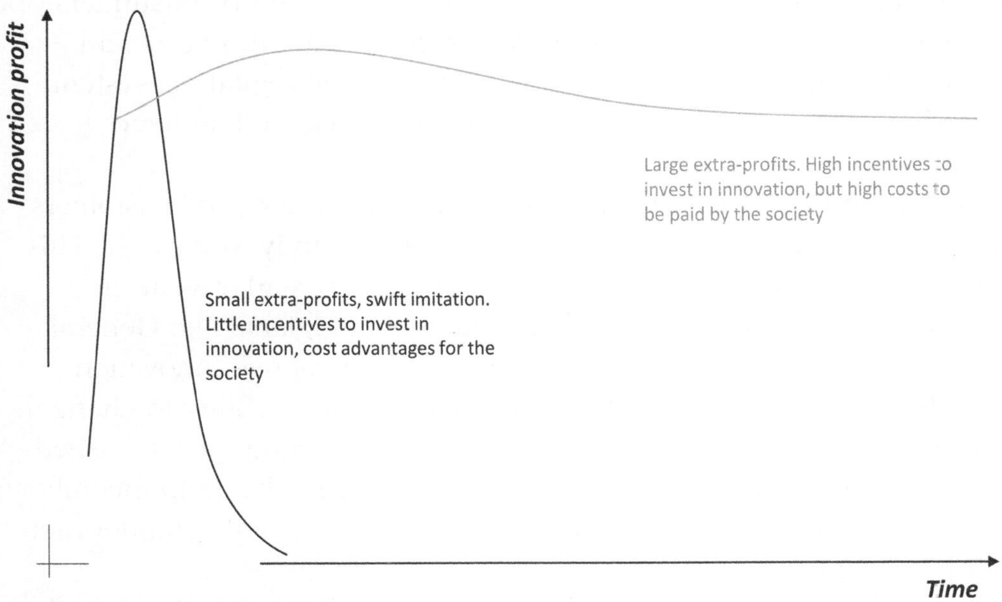

Figure 2.3 Schumpeter's innovation profit trade-off.

This is where institutional mechanisms come into play. Tools such as patents, regulatory frameworks, and antitrust policies act as levers that modulate the duration and intensity of temporary monopolies. They can either fortify a company's grip on a market or dismantle it to preserve fair competition. The case of the European Commission's fine against Microsoft for limiting access to interoperability codes is a striking example. Here, regulatory intervention was not about punishing success but about reestablishing balance, ensuring that no single player could leverage their dominance to block the emergence of others. Ultimately, the longevity and impact of a firm's market power depend as much on regulatory design as on the originality of its innovation.

Large, well-capitalized firms often invest heavily in internal R&D, building laboratories and innovation teams to chase the next big leap. Their scale offers advantages: more funding, lower capital costs, and access to talent. But with size comes inertia. These organizations can suffer from organizational slack—a creeping rigidity that slows their ability to respond to change despite their vast resources. In contrast, more agile entrants, often operating on thinner margins, can pivot quickly and challenge incumbents by disrupting the status quo.

This brings us to the concept of disruptive innovation, as shaped by Clayton Christensen. Disruption comes in two primary forms. In one scenario, *new market disruption*, innovators create entirely fresh demand by solving problems that had been previously ignored or misunderstood. Amazon's acquisition of Twitch in 2014 serves as a telling case, an investment that enabled the creation of a dynamic digital ecosystem around live-streaming, capturing a previously unserved audience segment.

The second form, *low-end disruption*, targets over-served customers with products that are "good enough" at a significantly lower cost. This was the case with Toyota's entry into the U.S. auto market with the affordable Corona. Toyota redefined value, forcing giants like General Motors to rethink their strategies. What makes disruptive innovation powerful is not just the novelty of the solution but its ability to change expectations, shift value, and challenge the very identity of established players. Ironically, the same organizational strengths that help incumbents succeed—like stable processes and strong values—can also hinder their ability to adapt.

But market dynamics are not defined solely by disruption. There is also a quieter, more strategic path—one that creates entirely new spaces rather than

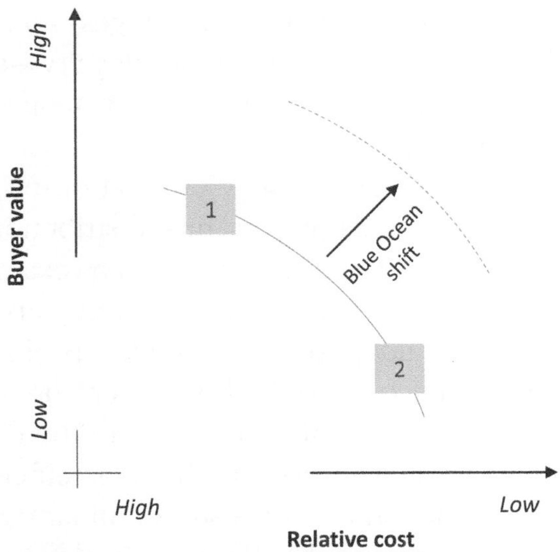

Figure 2.4 From market competing to market creating.

displacing old ones. This is the essence of the Blue Ocean Strategy (BOS), introduced by Chan Kim and Mauborgne (2005). Rather than battling for dominance in crowded, cutthroat markets—"red oceans"—BOS advocates for the creation of uncontested market space. As illustrated in Figure 2.4, this approach involves shifting away from the classic trade-off between cost (point 2 in Figure 2.4) and differentiation (point 1 in Figure 2.4). It calls on companies to reimagine the boundaries of their industry, unlocking new demand by offering products or services that redefine value altogether (see also section "Value Innovation" in Chapter 3).

In blue oceans, companies don't compete; they render competition irrelevant. Value innovation becomes the key: the simultaneous pursuit of differentiation and low cost to create leapfrog offerings that appeal to non-customers. These strategies challenge assumptions. They ask firms to think beyond market share and pricing wars to explore unmet needs, to reinvent customer experience, and to design for possibility rather than constraint. Industries from entertainment to aviation have applied this thinking, and many have found new paths to sustainable growth and competitive advantage.

Together, the dynamics of temporary monopolies, disruptive forces, and blue ocean creation reveal the rich complexity of modern market strategy. The challenge for today's leaders is not simply to innovate but to understand the strategic landscape in which innovation occurs.

Being a first mover can confer immense advantages—access to scarce resources, brand recognition, and early adopter loyalty—but it is not without its risks (Schumpeter, 1942; Porter, 1985; Christensen & Overdorf, 2000; Chan Kim & Mauborgne, 2005).

When a firm is first to introduce a new product or technology, it can secure valuable real estate: regulatory permits, distribution channels, supplier relationships, and customer mindshare. For example, the first fast food chain to dominate a market often benefits from superior logistics and operational scale. Early adoption also breeds customer lock-in, particularly when switching costs are high whether due to infrastructure, complementary goods, or user familiarity. Consider the enduring success of the QWERTY keyboard: not necessarily the most efficient layout but the first to gain widespread traction and thus still dominant. Similarly, Intel's early entry into microprocessors helped cement its role as an industry standard.

History is full of stories where first movers leveraged early entry to shape market trajectories, by carving path dependency advantages. Coca-Cola, launched in 1886, cultivated a brand so iconic that it remained market leader long after Pepsi's entry in 1898. In digital services, Zoom's 2011 debut allowed it to capture enormous ground during the COVID-19 pandemic. And in the personal computing revolution, Apple's 1984 Macintosh introduced a new way of interacting with machines, setting the stage for an entire generation of customer expectations.

Yet being first is not always synonymous with winning. In fact, late entrants often benefit from the groundwork laid by pioneers. They can avoid early missteps, learn from user feedback, fine-tune their offerings, and adopt newer, more efficient technologies. They don't need to educate the market or build the supply chain from scratch. Instead, they optimize what has already been tested. Lyft, for example, followed Uber into the ride-sharing space but carved out a distinct identity by focusing on community values and customer experience.

This pattern repeats across industries. Netflix followed earlier video rental services but reimagined content distribution through streaming. Google, Amazon, TikTok, and Tesla have all disrupted established players by combining first-mover innovation with adaptive strategy—the ability to sense shifts, respond quickly, and refine their value propositions over time.

One challenge for first movers lies in the immaturity of the ecosystem. Often, the technology itself is ahead of its time, and the infrastructure needed to support it—resellers, suppliers, and standards—is not yet in place. As a result, late entrants may benefit from a better-developed supply chain, clearer customer preferences, and lower experimentation costs. Moreover, early innovators may find themselves burdened with educating the market, convincing skeptical consumers, or facing delays due to insufficient readiness.

Even when first movers successfully shape customer preferences and build early loyalty, their position is not guaranteed. Sometimes, they must retreat, recalibrate, or even exit if the market proves unready or the offering misaligned. But even then, their efforts often leave behind a trail—ideas, lessons, and technologies—that others can follow.

In the end, the interplay between first-mover advantage, disruption, and strategic positioning reflects a deeper truth: in dynamic markets, success comes not only from being early or different but also from being attuned; attuned to change, to customers, to timing, and to the delicate balance between leading boldly and adapting wisely.

Balancing Innovation Protection and Diffusion

In the landscape of modern business, organizations are constantly navigating a high-stakes balancing act: how to protect the integrity and competitive value of their innovation initiatives, while also ensuring these innovations are successfully diffused into markets at the right pace and scale. This is not a binary choice but a dynamic and often delicate strategic tension—one that demands continual adjustment and thoughtful trade-offs. Far from being a purely operational concern, this tension touches the very heart of innovation strategy.

At its core, the challenge lies in managing a series of competing priorities that influence the direction and sustainability of innovation efforts. Recent academic literature has helped sharpen our understanding of these tensions. Scholars have shown that leaders must constantly reconcile flexibility with focus, openness with protection, and intuition with analysis. Organizations are increasingly called upon to be nimble enough to seize emerging opportunities, while maintaining the strategic discipline necessary to stay aligned with long-term goals. This is a hard balance to strike, especially in fast-moving markets where the shelf life of a good idea is short, and imitation by competitors is swift and often inevitable (Bingham & McDonald, 2022).

Within this landscape, leaders must also make decisions that weigh evidence-based precision against creative foresight. While incremental innovations often rely on data, trend analysis, and user feedback, disruptive leaps tend to emerge from bold intuition and a willingness to embrace uncertainty. Rigidly relying on one at the expense of the other limits potential. The tension extends to sourcing strategy: organizations must balance their internal expertise and embedded knowledge with an openness to external perspectives and fresh insight. Similarly, they must walk the line between efficiency and adaptability, between novelty and familiarity, and between brand consistency and market experimentation. Innovation, after all, does not occur in isolation; it must be communicated, trusted, and adopted. And this only happens when a brand delivers not just newness but genuine relevance.

One of the most powerful frameworks to emerge in response to these challenges is the notion of "sacrifice calculus." Rather than treating strategy as an exercise in wishful thinking—attempting to optimize growth, profitability, and innovation all at once—this model urges organizations to confront trade-offs directly and transparently. Strategic clarity, in this view, emerges not from ambition but from disciplined prioritization. The inability to choose between competing goals often leads to paralysis, confusion, or diluted impact. When everything is important, nothing truly is.

The sacrifice calculus reframes the strategy process as one of explicit exchange. For instance, a company might willingly accept lower short-term profit margins if doing so allows it to capture market share in a critical growth segment. Alternatively, it may sacrifice some degree of customization in favor of scalable efficiency. The key is to make these decisions deliberately, not reactively. Tools like sensitivity analysis can assist in this process, helping organizations model different scenarios, test divergent assumptions, and understand the ripple effects of their choices across departments, markets, and time horizons (Seidler et al., 2022).

But trade-offs don't end with abstract planning; they must be operationalized through resource allocation. And this, too, is fraught with complexity. One of the most persistent barriers to innovation execution is organizational inertia—the tendency to allocate capital, talent, and attention based on legacy patterns rather than emerging opportunities. Companies that fail to reallocate resources dynamically tend to underperform over time. In contrast, firms that regularly redirect funding from underperforming areas to high-potential ventures outperform their peers, achieving higher long-term valuations over time.

Overcoming inertia, however, is no small feat. It requires not only strong leadership but also systems and routines that disrupt the status quo. Companies must set clear innovation goals, define decision rules that promote change, and commit to difficult but necessary trade-offs. Creating a flexible, agile decision-making environment means treating resource allocation not as a fixed plan but as a living portfolio, constantly reviewed and adjusted, much like an investor rebalances their financial assets in response to market signals. In this context, agility is not just a virtue; it is a strategic necessity (Hall et al., 2012).

Taken together, these scholarly insights make one thing clear: managing innovation is not about generating a constant stream of ideas; it's about shaping and stewarding those ideas through an intentional strategy of trade-offs. It requires organizations to set priorities with conviction, to measure the impact of competing objectives, and to reconfigure resources in ways that reflect the fluid demands of the market. This is how innovation becomes both protected and diffused, not just imagined in a lab or studio, but transformed into real, scalable, market-shaping outcomes.

In essence, the balance between protection and diffusion is not a static target; it is an evolving process, one that must be reevaluated as markets shift, technologies mature, and organizational goals evolve. Companies that fail to make hard choices and commit to focused investments will find themselves stuck, neither competitive nor innovative. But those that embrace the discipline of trade-offs can ensure that their innovation efforts are not only bold but also enduring, delivering value that resonates today and remains relevant tomorrow.

Balancing Speed, Credibility, and Competitive Space in MVP Design

The Minimum Viable Product (MVP) is often misunderstood as a "minimal" version of a product. In practice, it is a disciplined market entry strategy: a deliberate first release designed to test the most critical assumptions under real conditions, with the least irreversible investment. For startups and Innovation Driven Enterprises (IDEs), the MVP is not a shortcut around quality but a method for converting uncertainty into evidence, quickly and repeatedly, before time and capital are consumed by speculative engineering.

The core tension is straightforward, yet unforgiving. An MVP must be lean enough to preserve speed, focus, and optionality but solid enough to function as a credible launching pad. "Lean" protects the venture from overbuilding in the wrong direction; "credible" protects it from being dismissed as incomplete, unreliable, or irrelevant. If the product feels too skeletal, the market may interpret it as an experiment rather than an offering—and early users, partners, and investors may hesitate to engage. If it is too ambitious, it can turn learning cycles into long development arcs that collapse the very advantage IDEs seek: rapid iteration.

What makes the decision particularly strategic is that MVP scope is also a competitive statement. A well-calibrated MVP signals that the team understands the problem deeply and is capable of delivering a usable solution, even under constraints. That signal matters because early markets are often defined as much by perception as by performance. When the first release demonstrates a coherent value proposition, it reduces ambiguity for customers and narrows the narrative space that competitors can exploit. Conversely, an MVP that looks like a prototype can unintentionally invite rivals to position themselves as the "serious" alternative and capture attention while the original entrant is still iterating.

This is why the MVP should be built around a sharp value hypothesis, not a feature checklist. The question is not "what can we ship fastest" but "what must be true for this business to work." In that sense, the MVP is a focused instrument: it tests the riskiest assumptions about user behavior, willingness to adopt, and the practical feasibility of delivery, onboarding, and retention. The right MVP is the one that produces decisive learning—the kind that validates a direction or forces a pivot—without requiring a level of investment that makes changing course psychologically or financially difficult.

At the same time, IDEs must treat "minimal" as a boundary, not a target. The first release must reach a threshold of usability and trust, especially when the product touches workflows, payments, data, safety, or professional identity. In these contexts, a weak MVP does not simply generate neutral feedback; it can damage credibility and slow adoption for subsequent iterations because early impressions are sticky. The discipline, therefore, is to strip down scope while protecting the user experience at the moment that matters most: the first encounter with the promised value.

The most successful MVP stories illustrate this principle through practical constraint, not theoretical elegance. Zappos validated demand for buying shoes online before building a full logistics engine; Groupon proved the appetite for deals using manual processes before automation; Dropbox reduced uncertainty through a simple demonstration that triggered measurable sign-ups; Twitter began as an internal utility before becoming a public platform. The common logic is consistent: early versions were not "small" for its own sake but intentionally designed to test adoption and engagement, while postponing expensive infrastructure until the evidence justified it.

Ultimately, the MVP is where strategy meets execution under uncertainty. Startups and IDEs succeed when they deliver an early version that is lightweight in cost and commitment yet substantial enough to claim a legitimate position in the market. The aim is to enter fast, learn faster, and still arrive with a product that closes the door behind it—not by building everything at once but by reaching a minimum threshold that customers respect and competitors must respond to.

References

Bingham, C. B., & McDonald, R. M. (2022). Mastering Innovation's Toughest Trade-Offs. *MIT Sloan Management Review.*

Chan Kim, W., & Mauborgne, R. (2005). Value innovation: A leap into the blue ocean. *Journal of Business Strategy, 26(4)*, 22–28.

Christensen, C. M., & Overdorf, M. (2000). Meeting the Challenge of Disruptive Change. *Harvard Business Review.*

Drucker, P. (2012). The Practice of Management. *Routledge* (first published in 1954).

Gallo, A. (2017). A Refresher on Discovery-Driven Planning. *Harvard Business Review.*

Girotra, K., & Netessine, S. (2014). Four Paths to Business Model Innovation. *Harvard Business Review.*

Hall, S., Lovallo, D., & Musters, R. (2012). Put Your Money Where Your Strategy Is. *McKinsey & Company.*

ISO (2020). ISO 56000:2020 – Innovation management – Fundamentals and vocabulary. *International Organization for Standardization.*

ISO (2024). ISO 56001:2024 – Innovation management system – Requirements. *International Organization for Standardization.*

McGrath, R., & McManus, R. (2020). Discovery-Driven Digital Transformation. *Harvard Business Review.*

Pisano, G. P. (2015). You Need an Innovation Strategy. *Harvard Business Review.*

Porter, M. E. (1985). Competitive Advantage. *Free Press.*

Reeves, M., Love, C., & Tillmanns, P. (2012). Your Strategy Needs a Strategy. *Harvard Business Review.*

Schumpeter, J. A. (1942). Capitalism, Socialism and Democracy. *Harper & Brothers.*

Seidler, M., Methling, F., & von Nitzsch, R. (2022). How to Make Strategic Trade-Offs. *Harvard Business Review.*

Thiel, P., & Masters, B. (2014). Zero to One: Notes on Startups, or How to Build the Future. *Crown Currency.*

Chapter 3

Exploiting Insights and Realizing Value: A Strategic Framework for Innovation Management

Systematic Exploitation of Insights and Value Realization According to the ISO 56000 Series

At the heart of the International Organization for Standardization (ISO) 56000 series lies a compelling vision: that innovation is not an isolated event but a structured, systemic discipline—one grounded in knowledge, shaped by context, and ultimately geared toward meaningful value creation. This series of international standards invites organizations to move beyond intuition or improvisation and instead build a formal innovation management system capable of harnessing insight and converting it into sustained advantage.

According to ISO 56001 (ISO, 2024), innovation begins with the capacity to listen, learn, and connect. As stated in *Clause 4.3.5.1*, organizations must draw on all available internal and external knowledge sources—whether from customers, partners, market signals, or scientific research—to expand their understanding of emerging needs. But this knowledge is not collected for its own sake. It serves a higher purpose: to inform, shape, and accelerate innovation efforts. The subsequent clause (*4.3.5.2*) highlights that developing effective, relevant solutions requires a systematic approach to identifying needs.

DOI: 10.4324/9781003743583-3

This means actively engaging with shifts in the operational context, tracking evolving expectations, and cultivating a mindset that thrives not on certainty but on intelligent responsiveness.

Crucially, ISO frames insight not as a static asset but as a strategic capability—one that must be prioritized, managed, and continuously sharpened. In *Clause 4.3.5.3*, it is made clear that organizations should align their long-term innovation strategy with their knowledge strategy. Doing so leads to better user satisfaction, faster time-to-value, stronger returns on innovation investments, and longer life cycles for new solutions. The goal is not just to be innovative but to be systematically insightful—to know more, sooner, and to act more effectively because of it.

But knowledge alone is not enough. Insights must be made actionable. In *Clause 4.3.5.4*, ISO 56001 outlines the operational requirements for turning information into value. This includes sourcing insights across disciplines, engaging directly with stakeholders, and building secure, accessible systems for managing and sharing knowledge. Organizations are expected to develop robust processes for identifying, analyzing, and using information, all while balancing the limits of available resources. Insight becomes a lever—not just for ideation but for innovation activities at large.

The realization of value—whether financial, social, environmental, or strategic—is positioned as the ultimate outcome of this system. As detailed in *Clause 4.3.1.1*, value is not measured only in terms of revenue or profit but also in the adoption, implementation, and long-term impact of new or improved solutions. Understanding and addressing stakeholder needs becomes essential—not only for competitive relevance but also for broader organizational legitimacy and growth. Innovation management, under ISO, becomes a matter of purpose-driven performance. This requires organizations to establish clear objectives, track meaningful indicators, and evaluate outcomes across multiple value dimensions.

Supporting this vision is a broad infrastructure of resources, detailed in *Clause 7*. Here, the standard makes explicit that organizations must determine, allocate, and sustain the full spectrum of resources needed to fuel innovation: skilled people, time, knowledge, financial support, and enabling infrastructure. It's not just about having these inputs; it's about orchestrating them with intention. That means assembling diverse, cross-functional teams; securing dedicated time for exploration; capturing and circulating knowledge effectively; and maintaining physical and digital spaces that encourage collaboration. Financial planning must reflect the risk-reward dynamics of innovation, while skills development and clear communication are framed as continuous imperatives. From documentation and tools to the protection of intellectual property, the entire system must be designed to support innovation as a core organizational capability (*Clauses 7.1–7.8*).

Equally vital is the organization's capacity to understand the environment in which it operates. As outlined in *Clauses 4.1 and 4.2*, this includes a comprehensive, ongoing assessment of external trends—economic, technological, and environmental—and internal conditions such as strategic direction, competencies, performance, and culture. This analysis helps define the scope of the innovation management system and ensures that it remains aligned with both internal aspirations and external realities. Within this frame, *Clause 4.4* introduces the concept of innovation intent—a forward-looking orientation that serves as the foundation for designing, using, and refining the system. In an age of heightened uncertainty, this becomes not just a strategic document but a navigational tool for innovation leadership.

Ultimately, building a robust innovation management system means weaving insight into the very culture of the organization. It's about more than process; it's about embedding learning, foresight, and agility into how the organization thinks and acts. The ISO 56000 (ISO, 2020) series provides the scaffolding for this transformation. It urges companies to bring together strategic ambition and operational discipline, to foster innovation not just as an outcome but as an ongoing commitment. Only by doing so can organizations hope to thrive in a global landscape marked by constant change and intense competitive pressure.

In this light, innovation is no longer a reactive effort. It becomes a structured, proactive journey—powered by insight, sustained by value, and enabled by a system that is as intentional as it is adaptive.

Mapping the Innovation Landscape: From the Solution-Problem Space to Utility-Meaning Models

In the evolving discipline of innovation management, a powerful analytical lens is the mapping of the interplay between a problem and its corresponding solution. This conceptual exercise reveals not only the degree of novelty embedded in an innovation but also its

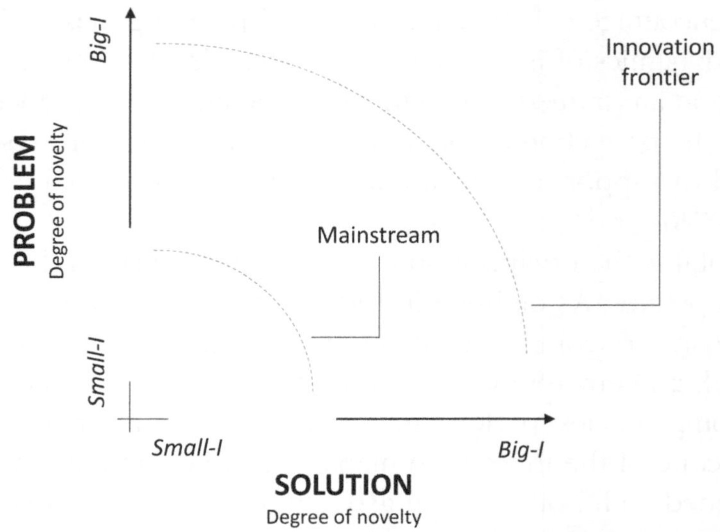

Figure 3.1 The Solution–Problem Innovation Space.

potential significance and impact. As illustrated in Figure 3.1, the Solution-Problem Innovation Space is structured around a two-dimensional grid. On the horizontal axis lies the solution's level of novelty and refinement, while the vertical axis captures the complexity or novelty of the problem being addressed.

In the bottom-left corner of this space—near the origin—both the problem and the solution are well understood and aligned with existing organizational practices or technological capabilities. Innovations found here tend to be incremental, often referred to as Small-I innovations. These are the realm of routine improvements: tweaks to existing processes, enhancements to operational workflows, and minor refinements that optimize performance without fundamentally disrupting the status quo.

However, as one moves outward from the origin—particularly diagonally across the space—one enters the realm of deeper divergence: the terrain where both the nature of the problem and the proposed solution deviate significantly from conventional models. This is the so-called innovation frontier, where radical or Big-I innovations emerge. These are the outcomes of strategic foresight, visionary research and development (R&D), design thinking, and often cross-sector fertilization. Big-I innovations typically solve complex or poorly articulated problems using breakthrough technologies or unorthodox combinations of knowledge. The quality and transformational potential of an innovation, in this framework, is seen as directly proportional to the quality of the problem-solution fit. That is, the most meaningful innovations occur not merely because they are novel but because they provide an effective response to needs that were previously unmet, misframed, or inadequately understood.

Beyond mapping novelty and fit, another critical facet of innovation lies in how solutions are interpreted, experienced, and valued by users. To make this tangible, innovation studies increasingly turn to structured approaches that connect a solution's attributes (i.e., its physical and functional features) to the benefits they enable, and ultimately to the core values held by consumers or stakeholders.

These connections are not always linear or obvious. For instance, an ergonomic design may offer comfort (a benefit) that resonates with a user's value for well-being. Or an eco-friendly material might not only reduce carbon emissions but also speak to a user's sense of ethical responsibility. By using tools such as laddering interviews and value hierarchy models, organizations can uncover these layered relationships and shift their focus from a purely product-centric approach toward a genuinely consumer-centric value proposition. This transition is essential for crafting innovations that not only function well but also mean something in the eyes of users.

This brings us to Figure 3.2, which presents two complementary models for evaluating technological solutions. The first model adopts a layered reading across three levels: the innermost layer, concerned with the "how" of technology—its architecture, materials, algorithms, and engineering logic; the middle layer, which focuses on the "what"—the functions and outputs that the technology delivers; and the outermost layer, where we consider the "why"—that is, the motivations, expectations, and performances shaped by user experience and interaction.

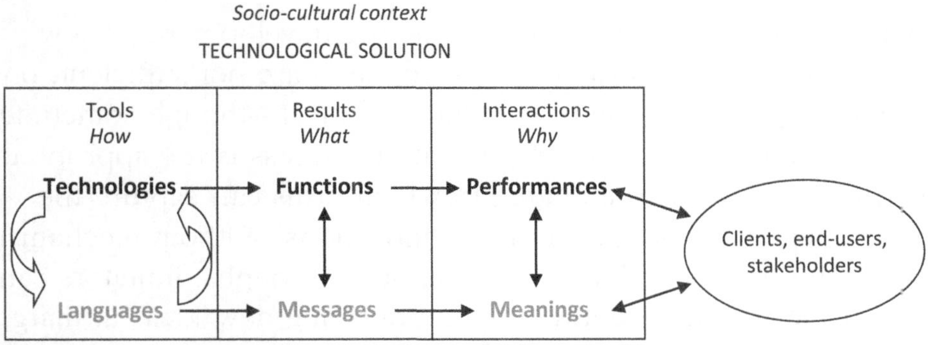

Figure 3.2 The utility-meaning framework.

The second model reframes this analysis through the twin dimensions of utility and meaning. The utility dimension captures the technical, performance-oriented side of innovation: speed, reliability, precision, and so forth. These are typically supply-side improvements. The meaning dimension, by contrast, deals with how users interpret and assign significance to those innovations. This is a socially constructed layer, shaped by perceived values (such as status, identity, or sustainability), corporate messages (what the company claims or implies), and the languages used to convey these messages (through design, naming conventions, packaging, or seasonal styling).

When organizations combine the problem-solution space of Figure 3.1 with the utility-meaning framework of Figure 3.2, they gain a richer, multidimensional perspective on innovation. Not only can they evaluate a solution based on its technical rigor and functionality but they can also assess its resonance with stakeholder expectations, values, and culture. This fusion of analytical tools supports more deliberate, user-aligned innovation strategies.

Ultimately, the act of innovating is not confined to engineering new tools or optimizing performance; it is about delivering offerings that feel meaningful and relevant in their socio-cultural and functional contexts. Through these integrated models, organizations are better equipped to evaluate, refine, and position their innovations with strategic foresight and empathic insight (Vriens & Hofstede, 2000).

Appropriability Regime and Value Innovation

The Appropriability Regime and Legal Protection

In contemporary innovation ecosystems, the development of new products and the pursuit of advanced research are not sufficient, on their own, to generate long-term value. A critical—though sometimes underestimated—component of innovation success is the appropriability regime, which governs the degree to which firms can capture the economic benefits arising from their innovations. Without mechanisms to protect and extract value from these breakthroughs, imitators can quickly erode competitive advantage, replicating new ideas at marginal cost and depriving the original innovators of returns on their R&D investments.

Appropriability regimes encompass a blend of institutional, legal, and resource-based conditions that influence how the benefits of innovation are distributed across stakeholders—innovators, customers, suppliers, and competitors alike. When this regime is strong, it enables the originator to appropriate a significant portion of the value created. Conversely, when it is weak, the gains tend to disperse more broadly through the market, often to the benefit of fast followers or larger actors with better access to distribution or scaling resources.

Four primary elements shape the effectiveness of an appropriability regime. The first concerns intellectual property rights (IPR)—patents, copyrights, trademarks, and trade secrets. These legal instruments are foundational in protecting innovations, but their effectiveness is highly context-dependent. For instance, patents offer robust protection in domains such as pharmaceuticals or mechanical engineering, where novelty is clear and technical documentation does not easily enable replication. Yet in areas like software or advanced manufacturing, patents may provide limited deterrence due to the ease with which others can design around them or reinterpret their scope. In such cases, companies may strategically opt for secrecy as their primary defense, especially when public disclosure might facilitate imitation. This trade-off between protection and disclosure is a strategic calculation, not merely a legal matter (See also Chapter 2, paragraph 2.5.2 "Balancing Innovation Protection and Diffusion").

The second determinant lies in the availability and control of complementary assets. These may include financing, supply chain capabilities, production know-how, or access to key marketing channels. When these assets are held in-house, innovators are better positioned to capture value. However, when they must rely on external parties, issues of bargaining power emerge. The more specialized these complementary assets are, the greater the interdependence—and the more complex the value-sharing arrangement becomes. On the other hand, when assets are generic and widely available, they pose fewer obstacles to innovation diffusion but also dilute the originator's power to appropriate returns.

Third, the imitability of the innovation plays a vital role. This is shaped not only by legal protection but also by the nature of the knowledge underpinning the innovation, whether it is tacit or codified. Codified knowledge—which can be documented, taught, or transmitted—tends to be easier to replicate unless tightly protected. In contrast, tacit or "sticky" knowledge, deeply embedded in routines, culture, and experiential learning, serves as a more elusive barrier to entry. This kind of knowledge often proves harder for competitors to emulate, thus offering a transient yet valuable shield.

The fourth element concerns timing. When firms are early to market, they can exploit a time-based advantage to establish brand recognition, build customer loyalty, and iterate quickly. Pioneering companies like Microsoft, Intel, and Motorola have historically leveraged these windows by reinvesting profits into new product development, thereby reinforcing their competitive position. Such cycles of fast learning and continuous improvement, especially in sectors governed by steep experience curves, can be critical in consolidating early leads.

This tension between incentivizing innovation and ensuring societal benefit is captured effectively in Figure 3.3, which illustrates the trade-off between monopolistic profits and collective welfare. When a company successfully innovates—say, through a cost-reducing process innovation—it might set the price at a level (P = c) corresponding to the pre-innovation unit cost, represented by point c. Following its innovation investment of Δ, the firm could, under temporary monopoly rights, raise prices and capture monopoly profits (area M). The total collective surplus thus rises from CS (consumer surplus) to CS + M. However, there is a social loss (SL)—a portion of the potential welfare not realized due to constrained pricing and output. If those temporary rights were hypothetically revoked post-disclosure, prices might fall to (c – Δ), maximizing the collective benefit (CS + M + SL). But in doing so, society would undermine the *ex ante* incentive for firms to invest in risky innovation in the first place—a paradox that underscores the fragile balance between private return and public gain.

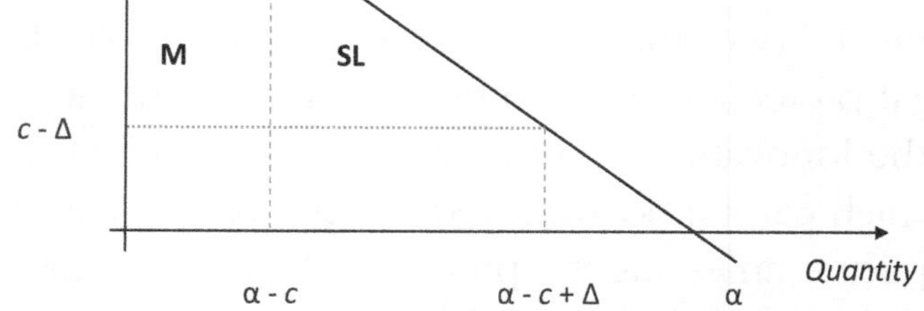

Figure 3.3 Monopolistic profits vs. societal benefits trade-off.

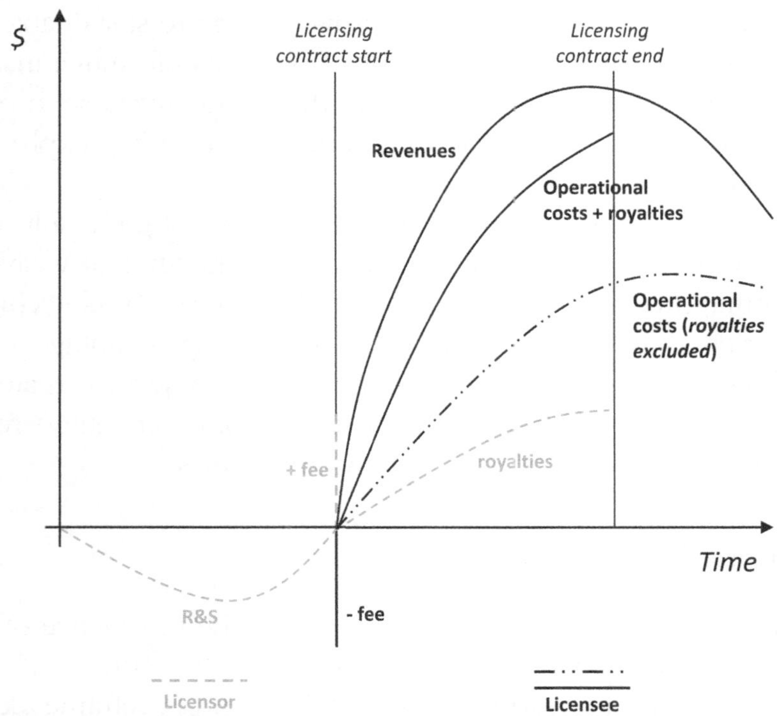

Figure 3.4 Licensor vs. licensee revenues-costs structure over time.

Figure 3.4 provides a complementary perspective by mapping the revenue-cost dynamics in a licensing scenario, contrasting the licensor—the originator of the innovation—with the licensee, who gains the rights to commercialize it. Initially, the licensor absorbs the brunt of R&D costs, sinking capital before any returns materialize. This financial effort is later offset through an upfront licensing fee and ongoing royalty payments, creating a stable revenue stream over time. For the licensee, meanwhile, the licensing fee represents an initial investment that is gradually recouped as operational costs are surpassed by increasing revenues. However, while licensing accelerates market diffusion and enables monetization, it may also dilute competitive advantage, particularly if the same innovation is licensed to multiple players across geographies or sectors.

Choosing between patenting and secrecy is, in the end, a strategic decision grounded in the nature of the knowledge and the risk of imitation. Codified knowledge lends itself to patenting, where controlled disclosure doesn't significantly undermine competitive position. In contrast, when knowledge is largely tacit or "sticky"—embedded in processes, culture, or proprietary know-how—secrecy often provides a more sustainable barrier. One of the most iconic examples remains Coca-Cola's formula: instead of patenting it and revealing its composition, the company chose to preserve its mystique through long-term trade secrecy, thereby avoiding expiry, imitation, and commodification.

In conclusion, a robust appropriability regime is not the result of legal safeguards alone. It is the product of a coherent strategy that weaves together formal protections, control of complementary assets, technology-specific barriers to imitation, and time-based learning advantages. When aligned effectively, this system enables innovators not just to create value but to retain a meaningful share of it, ensuring both competitive returns and the financial sustainability of future innovation efforts.

Value Innovation

A recurring obstacle in the innovation process is the prevalence of mental rigidity and conventional assumptions. These cognitive constraints inhibit the ability of organizations to reimagine their offerings, reframe existing challenges, or perceive latent opportunities. One effective tool to counter this inertia is the *principle of subtraction*, a deliberate method that involves temporarily removing key components from a product, service, or process to simulate an alternative version of the original. This subtractive modeling reveals hidden advantages that are often overlooked, ranging from simplified design and cost reductions to increased usability and even enhanced environmental performance (Kumar, 2012).

This method does not simply aim at eliminating; rather, it allows for an evaluative exploration of what value remains, what emerges in its place, and whether this altered configuration can serve new market segments. By systematically iterating this approach across multiple components, organizations heighten the probability of uncovering discontinuous innovations—those that break away from incremental paths and redefine industry norms (Almquist et al., 2016).

Such thinking forms the cognitive and strategic basis for what has become known as the Blue Ocean Strategy (BOS), developed by Chan Kim and Mauborgne (2005). Traditional business environments—termed "red oceans"—are characterized by well-defined market boundaries, saturated competition, and the zero-sum logic of either cost leadership or differentiation. Within this framework, companies are locked in competitive combat over shrinking margins and static demand.

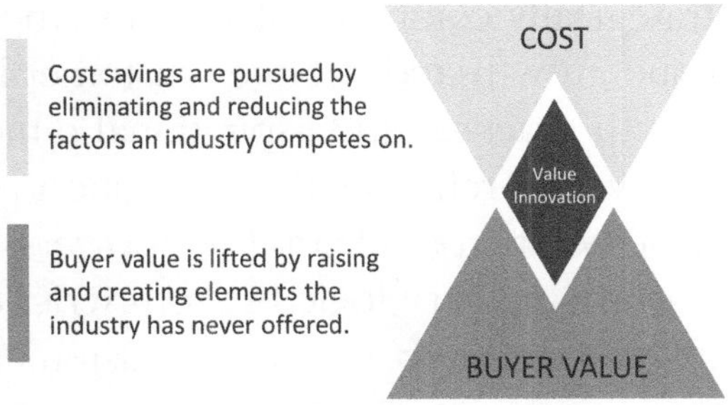

Figure 3.5 Value innovation.

In contrast, BOS proposes a radically different logic: organizations should aim to generate "blue oceans" uncontested market spaces where competition becomes irrelevant because demand is created rather than fought over. At the core of this philosophy lies the concept of value innovation, where companies simultaneously pursue low-cost and high-value creation, thereby disrupting the traditional value-cost trade-off (see also section "Disruptive Change and First-Mover Advantage" Chapter 2).

Figure 3.5 visually articulates this principle. The inverted triangle structure shows how cost can be reduced by eliminating or minimizing industry-standard factors, while buyer value is elevated by raising and creating elements that the industry has never previously offered. The convergence of these efforts in the central zone of "value innovation" marks the sweet spot where strategic differentiation and operational efficiency are achieved simultaneously. This model challenges the prevailing assumption that competitive advantage must come either from cost or uniqueness; it argues for both.

Value innovation thus encourages organizations to redesign their offerings, not simply by enhancing features, but by reconfiguring the value landscape. This includes questioning which industry norms are outdated, which customer expectations are under-addressed, and which functional redundancies may be pruned to streamline performance and cost.

To deepen this understanding, the Elements of Value Framework proposed by Almquist et al. (2016) introduces a layered perspective. It categorizes 30 value elements into four ascending levels: functional (e.g., cost savings, product quality, time efficiency), emotional (e.g., reduced anxiety, design aesthetics), life-changing (e.g., personal growth, motivation), and social impact (e.g., contribution to community or environmental sustainability).

Companies that strategically combine value across these levels tend to achieve faster revenue growth and stronger customer loyalty, as their offerings resonate on both practical and aspirational dimensions.

Ultimately, value innovation reframes the innovation process from a feature-centric to a value-centric paradigm. It empowers firms not merely to differentiate within existing industry logic but to rewrite the rules altogether, anchoring their advantage in new spaces where customer needs are redefined—and, in many cases, newly created.

Innovation Appropriability and Value Distribution

When an innovation is introduced into the market, its value is rarely confined to the boundaries of the firm that created it. Instead, this value becomes dispersed across a web of actors, including competitors, suppliers, users, and, of course, the innovator itself. The extent to which the innovating firm can capture a significant portion of this value hinges on what scholars refer to as the appropriability regime: a configuration of legal, institutional, and resource-based mechanisms that either protect or expose innovation-derived profits (Teece et al., 1997).

To better understand how this dynamic unfolds, we can turn to the conceptual matrix illustrated in Figure 3.6, which categorizes innovation environments along two dimensions. The first is the strength of the appropriability regime, ranging from weak to strong. The second is the nature of complementary assets—those supporting capabilities or infrastructures (such as distribution, branding, and manufacturing know-how) that can either be freely available or tightly held. The intersection of these two variables results in four strategic configurations, each with distinct implications for value capture.

In the first quadrant, where appropriability is weak and complementary assets are freely accessible, the innovator is at a distinct disadvantage.

	Free-Available Complementary Assets	Tightly-Held Complementary Assets
Weak Appropriability Regime	Niche High-Tech Products (e.g., underwater cameras, 3D printers)	Innovative Consumer Goods (e.g., smart wearables with unique patents)
Strong Appropriability Regime	Open Software Platforms (e.g., Android OS with strong community but proprietary control)	Biopharmaceuticals (e.g., patented drugs requiring clinical trials and distribution networks)

Figure 3.6 Matrix of innovation appropriability and complementary assets.

Here, the barriers to imitation are low, and competitors can easily replicate both the innovation and the means to commercialize it. Products like niche high-tech gadgets—such as underwater cameras or certain 3D printers—often fall into this category. Despite their technical sophistication, they suffer from limited legal protection and rely on widely accessible components or processes. The result is a market where value is diffused, price competition is fierce, and the innovator struggles to secure sustainable returns.

The second quadrant maintains the weak appropriability context but combines it with tightly held complementary assets. This changes the calculus. Even if intellectual property protections are minimal, the fact that certain crucial resources—like specialized suppliers or proprietary production techniques—are not easily replicated acts as a deterrent to imitation. A relevant example can be found in innovative consumer goods, where the uniqueness of production often resides more in the manufacturing network than in the design itself. This enables better value capture despite the absence of strong IPR defenses.

Moving into the third quadrant, we encounter a scenario characterized by strong appropriability regimes paired with freely available complementary assets. This is typical in sectors like software, where legal tools such as copyrights, patents, or proprietary licensing structures provide robust protection, even though development environments and supporting technologies are often open-source or widely distributed. A case in point is the Android operating system, which benefits from a large, collaborative community while retaining proprietary control mechanisms that preserve strategic advantage for its primary custodians.

Finally, in the fourth quadrant, both appropriability and control over complementary assets are high—creating the optimal conditions for capturing value. This is the domain of industries like pharmaceuticals, where patented drugs not only enjoy legal exclusivity but also require highly specialized distribution networks, clinical trial infrastructures, and regulatory expertise. These barriers create a fortress-like structure around the innovation, ensuring that the lion's share of the value remains with the innovator.

As Figure 3.6 illustrates, these four archetypes offer more than mere classification; they serve as strategic lenses. For instance, the relatively low market price of underwater cameras can be explained by their positioning in the weak–free quadrant: replication is easy, and protections are scarce, eroding the innovator's ability to retain value. Conversely, the high margins enjoyed by biotech firms stem from the robust protections and exclusive infrastructures they command.

Ultimately, understanding this matrix empowers innovators to make more deliberate strategic choices not only regarding their technological developments but also in how they manage, acquire, or safeguard the complementary assets that surround them. Designing innovation strategies without considering this ecosystem of appropriability and asset control risks diluting competitive advantage, while aligning both dimensions can pave the way toward sustained value appropriation and defensible market leadership.

Free Cash Flow as a Measure of Value Creation

When evaluating a company's capacity to generate and sustain value, one must carefully distinguish between the lenses of accounting and finance. Though interdependent, these two disciplines offer different, often complementary, views on economic performance. Accounting, with its backward-looking orientation, emphasizes historical data, focusing on metrics such as net income and standardized reporting structures. Finance, by contrast, projects forward, orienting itself around the long-term generation of free cash flow (FCF)—a key indicator of a business's economic vitality and scalability.

This distinction is particularly salient in the context of startups, where traditional accounting measures can obscure a venture's true financial health. In the early phases of a startup's lifecycle, net income is often negative due to aggressive investment in growth and innovation. Yet, this does not necessarily signal a lack of value creation. What matters more is the firm's ability to generate FCF over time—cash that remains after covering operational expenditures and capital investments. This surplus can be reinvested in the business, distributed to shareholders, or used to service debt. As such, FCF serves as a more telling indicator of a venture's long-term economic promise.

Figure 3.7 provides a breakdown of the FCF calculation. The process begins with Earnings Before Interest and Taxes (EBIT). From this, taxes are deducted to arrive at Earnings Before Interest and Taxes After Taxes (EBIAT). Next, non-cash expenses—most notably depreciation and amortization—are added back since they reduce accounting profits but not actual cash flow. Adjustments for changes in net working capital are then

Free Cash Flow Equation

EBIT
- Taxes
EBIAT
+ Depreciation and Amortization
+/- Changes in Working Capital
- Capital Expenditures

FREE CASH FLOW

Figure 3.7 Calculation of free cash flow.

applied, reflecting the firm's short-term operational liquidity. Finally, capital expenditures (CapEx) are subtracted, capturing reinvestments necessary to maintain or grow productive capacity. The result is FCF: the lifeblood of financial valuation and strategic investment.

In the financial realm, FCF is often assessed through the prism of discounted cash flow (DCF) analysis. Here, the projected stream of future FCFs is discounted at a rate known as the weighted average cost of capital (WACC), a composite of the firm's equity and debt costs. If the expected returns from these cash flows exceed the cost of capital, the firm achieves a positive net present value (NPV). This signals that the investment is not only viable but also value-creating from an economic standpoint.

However, generating FCF in excess of WACC is only the beginning. The real challenge lies in sustaining this surplus overtime, a task that requires strategic foresight and resilience. Firms must continuously invest in innovation, protect intellectual property, and erect barriers to entry such as strong brand equity or proprietary technologies. These elements help fortify a company's competitive moat, shielding its market position from imitation and erosion. At the same time, scalable growth plays a pivotal role. As operations expand, so does the base of cash flows, compounding the firm's ability to create enduring economic value.

In this sense, FCF is not merely an accounting output but a strategic input—one that reflects both past decisions and future potential. It embodies the interaction between growth, risk, reinvestment, and protection of strategic assets. For founders and investors alike, this metric offers a dynamic lens through which to assess whether a venture is not only profitable in accounting terms but also economically sustainable in the long term.

Thus, while accounting delivers a standardized snapshot of past performance, finance provides the narrative arc of a company's journey, anchored by FCF and the ever-present benchmark of the cost of capital. Success, ultimately, lies in consistently generating returns that exceed this threshold, while building structural advantages that endure in an unpredictable, competitive landscape.

Uncovering Deep Insights and the Cultural Base of Innovation

Systematic Insight Generation

At the heart of any enduring innovation strategy lies not just creativity or bold experimentation but insight—not just data or ideas but meaningful, actionable understanding. Insights are the quiet engine behind breakthrough innovations—the subtle revelations that connect unmet customer needs with emerging technologies, overlooked market signals, or hidden organizational frictions. Without them, innovation becomes a guessing game. It's no wonder, then, that industries often report very high failure rates when ideas are launched without true insight at their core.

True insight is not accidental. It must be deliberately cultivated through a systematic process of exploration and synthesis. One proven model for doing so is a five-phase cycle: *planning, collecting, connecting, selecting,* and *sharing.* While it may seem deceptively linear, this process is iterative and dynamic, fueling a culture of ongoing discovery.

In the planning phase, the team defines a clear objective and aligns on the scope of inquiry. What are we trying to understand? What assumptions need to be challenged? What signals are already surfacing? From there, in the collection phase, the organization enters an open-ended phase of immersion, tapping into sources as varied as trend analysis, expert interviews, ethnographic research, lead-user observations, and digital analytics. These are not simply exercises in data gathering; they are invitations to see the world differently, often requiring companies to cultivate new lenses for observation.

But insight only begins to take shape in the connect phase, where the real work of meaning-making begins. Here, diverse data points, sometimes seemingly unrelated, are juxtaposed, cross-referenced, and synthesized.

It's at this intersection of unexpected patterns where transformative insights tend to emerge, where a subtle shift in user behavior might reveal a gap in an existing service model, or where a weak signal from a new market combines with a technological anomaly to forecast a disruption. The selection phase then acts as a critical filter. Not every observation is valuable, and not every "interesting" insight is actionable. Leaders must prioritize those insights that align with strategic goals and exhibit the highest potential to unlock new value. Finally, in the sharing phase, insights are translated into narratives—stories that resonate across the organization, mobilizing support and anchoring the insight in collective memory.

This process, when done consistently, becomes more than a framework. It evolves into what has been described as an insight engine: a dynamic capability that transforms an organization's ability to perceive, understand, and act. Van den Driest et al. (2016) emphasized the power of such engines in companies like Unilever, where disciplined insight generation is embedded in decision-making at the highest levels. Insight engines are not powered by software alone; they depend on a marriage of analytical depth and human intuition, balancing the precision of data analytics with the subtleties of interpretation and judgment.

A particularly powerful component of this engine lies in an organization's sensitivity to anomalies—the outliers, the out-of-pattern behaviors, and the unexpected bumps in the data. Rather than dismissing these irregularities, high-performing innovators investigate them, recognizing that anomalies often signal unmet needs or underappreciated shifts in customer behavior. Martin et al. (2021) argue that cultivating a culture of curiosity and pattern disruption is essential for turning weak signals into first-mover advantage.

But anomalies are just one source. Sawhney and Khosla (2014) expand the field of vision with their framework of seven insight archetypes: *anomalies*, *confluences*, *frustrations*, *orthodoxies*, *extremities*, *journeys*, and *analogies*. Each of these lenses offers a pathway into deeper, more meaningful understanding. By asking, for instance, "What do our most extreme users do differently?" or "Which long-held industry belief no longer serves us?", companies shift from passive observation to strategic insight generation.

Taken together, these practices form more than a methodology. They reflect a cultural orientation—one that prizes insight not as a discrete output but as an embedded habit of mind. In this view, insight is not something found, but something built: through disciplined exploration, deliberate reframing, and the willingness to challenge what we think we know.

In a world increasingly driven by complexity and ambiguity, cultivating insight fluency may well be the most important organizational capability of all. When deep insights become the foundation of innovation, companies move beyond guesswork and serendipity, unlocking strategies that are not only imaginative but also profoundly grounded in what truly matters.

Customer-Centered Perspectives and Market Redefinition

In today's increasingly experience-driven economy, true innovation doesn't begin with the product; it begins with people. It starts not by asking, What can we build next? but rather, What are customers really trying to accomplish—and why does it matter to them? The shift toward customer-centric innovation challenges firms to move past traditional segmentation, past demographic guesses, and surface-level preferences, and instead focus on the actual problems customers are trying to solve in their lives.

At the heart of this approach lies the Customer-Centric Innovation Map, a framework introduced by Bettencourt and Ulwick (2008) that breaks down customer activity into six functional phases: *define goals*, *identify resources*, *prepare*, *execute*, *monitor*, and *modify*. These stages provide more than a workflow; they serve as a diagnostic lens for uncovering hidden frustrations, unmet needs, and moments of unnecessary complexity in a customer's journey. By analyzing each phase, organizations can discover where current solutions fall short and where the real opportunity for innovation lies.

This method encourages organizations to redefine what they mean by "value." Rather than layering on more features or adding bells and whistles, the goal is to eliminate friction, simplify decisions, and increase the chances of success for the customer. In short, innovation becomes an act of empathy designed not around products, but around human progress.

Take the now-iconic Swiffer example from Procter & Gamble. The company resisted the temptation to simply upgrade the mop. Instead, they zoomed out and re-examined the entire experience of home cleaning. They applied the Innovation Map to break down the cleaning task into stages: from defining the cleaning goal and gathering supplies to preparing the space, performing the task, checking the result, and refining the routine. Along the way, P&G discovered that customers weren't just trying to clean floors; they were trying to do it quickly, neatly, and without hassle. The conventional mop failed on multiple fronts: it was messy, time-consuming, and awkward to store. Swiffer's success lay not in technical superiority but in offering a frictionless experience that aligned with real-world needs.

A similar shift in perspective helped Spotify redefine how people consume music. Instead of focusing on music ownership—a model entrenched in decades of buying records, CDs, and downloads—Spotify embraced a deeper job-to-be-done (JTBD): "I want instant, personalized access to music that matches my mood, my moment, or my movement." They recognized that listeners didn't just want songs; they wanted soundtracks to their lives. By organizing content around moods, activities, and personalized playlists, Spotify positioned itself not as a music distributor, but as a context-aware companion. It's not about offering all music. It's about offering the right music, at the right moment, with minimal effort.

The JTBD theory, as developed by Christensen et al. (2016), takes this logic even further. JTBD suggests that customers don't just buy products; they hire them to do a specific job. This reframing forces companies to look beyond superficial reasons for purchasing and dig into the emotional, functional, and social motivations behind behavior.

One of the most cited JTBD examples comes from McDonald's, which learned that morning commuters were "hiring" milkshakes not just for flavor or price but as a convenient, long-lasting, one-handed breakfast to make car rides less monotonous. This insight radically changed how the product was positioned and improved, proving that even mature categories contain untapped insight when viewed through the right lens.

Another compelling example is Airbnb, whose meteoric rise can be traced back to a deep insight about travel. Customers weren't simply booking lodging; they were "hiring" travel experiences to feel connected, safe, and immersed in new cultures. Traditional hotels failed to satisfy this job for many travelers, offering sterile consistency but lacking authenticity. Airbnb's platform offered an alternative: the ability to "live like a local." The company didn't just disrupt the hospitality industry; it reframed it around a different set of customer needs, creating new value where incumbents saw no opportunity.

What unites these stories is a willingness to challenge legacy assumptions, a readiness to listen more deeply to customer struggles, and the capacity to design around purpose rather than product.

Companies are committed to embedding these customer-centered perspectives in their organizational culture. It requires humility, because customers often can't articulate their needs clearly. It demands curiosity because the most powerful insights often hide in anomalies and workarounds. And it requires the courage to act because true customer-centered innovation often means disrupting your own business before someone else does.

In addition, by integrating frameworks like the Customer-Centric Innovation Map and JTBD theory, organizations gain more than strategic clarity; they gain empathy at scale. They begin to see their markets not as static segments, but as living systems of goals, tensions, and opportunities. In doing so, they don't just improve products; they create new categories, reshape expectations, and lead their industries forward.

References

Almquist, E., Senior, J., & Bloch, N. (2016). Elements of Value. *Harvard Business Review.*

Bettencourt, L. A., & Ulwick, A. W. (2008). The Customer-Centered Innovation Map. *Harvard Business Review.*

Chan Kim, W., & Mauborgne, R. (2005). Value innovation: A leap into the blue ocean. *Journal of Business Strategy, 26(4),* 22–28.

Christensen, C. M., Hall, T., Dillon, K., & Duncan, D. S. (2016). Know Your Customers' Jobs to Be Done. *Harvard Business Review.*

ISO (2020). ISO 56000:2020 – Innovation management – Fundamentals and vocabulary. *International Organization for Standardization.*

ISO (2024). ISO 56001:2024 – Innovation management system – Requirements. *International Organization for Standardization.*

Kumar, V. (2012). 101 Design Methods: A Structured Approach for Driving Innovation in Your Organization. *John Wiley & Sons.*

Martin, R., Goodson, B., & Kevin, W. (2021). The Power of Anomaly. *Harvard Business Review.*

Sawhney, M., & Khosla, S. (2014). Where to Look for Insight. *Harvard Business Review.*

Teece, D. J., Pisano, G., & Shuen, A. (1997). Dynamic capabilities and strategic management. *Strategic Management Journal, 18(7),* 509–533.

Van den Driest, F., Sthanunathan, S., & Weed, K. (2016). Building an Insights Engine. *Harvard Business Review.*

Vriens, M., & Hofstede, F. T. (2000). Linking attributes, benefits, and consumer values. *Marketing Research, 12(3),* 4–10.

Chapter 4

Adaptability and Managing Uncertainty

The Adaptability and Managing Uncertainty According to ISO 56000 Series Requirements

In a world where innovation is both a necessity and a moving target, adaptability and the ability to manage uncertainty stand out as two of the most vital principles for any organization pursuing sustainable growth. These principles are at the heart of the International Organization for Standardization (ISO) 56000 series, elaborated specifically in *ISO 56001* (ISO, 2024) and *ISO 56002* (ISO, 2019). But they are not abstract concepts; they are operational imperatives, deeply interlinked, shaping how an organization must think, plan, and act in the face of disruption, complexity, and fast-evolving opportunities.

According to *ISO 56000* (ISO, 2020), the principle of adaptability is not simply the ability to change but the capacity to anticipate and proactively reshape structures, competencies, and value creation models in response to evolving internal and external conditions. It's about responsiveness, yes, but even more, it's about resilience and foresight. As innovation unfolds, often in nonlinear or disruptive ways, the ability to recalibrate organizational systems becomes a decisive advantage, as emphasized in *Clause 4.3.7.1*.

This adaptive mindset acknowledges that innovative solutions rarely fit neatly into existing molds. They frequently require restructuring, not just in operations, but also in strategy, governance, talent deployment, and cultural norms. *ISO 56000* highlights that adaptability brings a range of benefits:

DOI: 10.4324/9781003743583-4

accelerated learning cycles, improved risk response, better use of resources, and more effective realization of innovation outcomes (*Clause 4.3.7.3*). Adaptability is not just reactive; it's the enabler of intelligent design amid uncertainty.

To embed adaptability meaningfully, *ISO 56002* proposes a series of practical actions, including assessing the organization's capacity to detect and interpret change, using scenario planning as a foresight tool, aligning resource allocation with anticipated disruptions, and building human capability through empowerment and learning. Organizations are encouraged to operate across multiple time horizons, manage competing initiatives in parallel, and set up responsive monitoring systems to govern change in real time (*Clause 4.3.7.4*).

Tightly interwoven with adaptability is the ISO principle of managing uncertainty. While adaptability speaks to how an organization morphs over time, managing uncertainty addresses its relationship with the unknown. Innovation, by nature, thrives in ambiguity. Rather than resisting it, *ISO 56000* urges organizations to embrace uncertainty as a condition of innovation and to manage it through structured risk balancing, ongoing experimentation, and learning loops, as outlined in *Clause 4.3.6.1*.

This approach calls for more than traditional risk management. It asks organizations to actively leverage uncertainty as a creative force. As *ISO 56000* notes, managing uncertainty is not about avoiding failure but developing confidence in one's systems and people to navigate the unknown, discover opportunity, and build resilience through insight (*Clause 4.3.6.2–3*).

ISO 56002 offers actionable guidance here as well. It recommends building robust uncertainty management frameworks, nurturing a culture that supports safe-to-fail experimentation, using portfolio-based approaches to balance risk and reward, and equipping teams with the tools, skills, and performance indicators needed to identify and respond to uncertainty at multiple levels (*Clause 4.3.6.4*).

These principles are not standalone; they are deeply embedded in the operational planning requirements of *ISO 56001*. For instance, *Clause 6.1* emphasizes the importance of planning actions to address both risks and opportunities in a way that integrates uncertainty considerations directly into the innovation management system. This includes setting clear objectives, defining thresholds for acceptable risk, and ensuring that all functions—from product design to go-to-market strategy—operate with a shared understanding of uncertainty.

Organizational structures must also reflect these priorities. *ISO 56001* recommends that structural configurations remain dynamic, especially when tackling radical innovations. This may involve dedicated teams or internal ventures that require different leadership styles, performance expectations, or resource flows than traditional operational units (*Clause 6.3*). Such differentiation is key when working across innovation horizons—from incremental improvements to long-term, high-risk bets.

To manage innovation portfolios effectively, *ISO 56001* suggests maintaining a healthy balance of initiatives with varied timeframes, risk profiles, and resource intensities (*Clause 6.4*). This kind of portfolio thinking not only buffers against uncertainty but also strengthens alignment with strategic priorities and market shifts (see Figure 1.3 "A Small-I to Big-I portfolio approach.").

Even the mechanisms of continuous improvement and corrective action—articulated in *Clause 10* of *ISO 56001*—are reframed through the lens of adaptability and uncertainty. Organizations are expected to treat every non-conformity not as a failure to be punished but as a learning opportunity that can feed a self-correcting, ever-evolving innovation ecosystem. Timely identification of root causes, swift corrective actions, and open documentation of lessons learned all contribute to a more agile and forward-looking organizational culture.

In sum, adaptability and uncertainty management are not peripheral to innovation; they are its backbone. These ISO principles call for an operating system that is both robust and responsive, capable of shifting gears without losing strategic coherence. When these capabilities are embedded into the DNA of the innovation management system, organizations are better equipped to seize emerging opportunities, mitigate risks intelligently, and sustain innovation performance over time. In a world defined by volatility, such capabilities are not optional—they are existential.

Experimental Innovation Loops: Reducing Uncertainty and Entrepreneurial Risk

As innovation continues to reshape industries at an accelerating pace, large corporations often find themselves at a strategic crossroads. Their structured, process-heavy models—designed for stability and scale—can inadvertently become roadblocks when navigating the volatile terrain of entrepreneurial innovation. Unlike Innovation-Driven Enterprises (IDEs), which are built to embrace uncertainty, many large firms have historically relied on a single, linear innovation cycle. These long, resource-intensive loops often push ideas forward through development and deployment without first validating the fundamental assumptions that underpin them. The result? Missed market signals, wasted investments, and, too often, products or services that arrive too late, or never truly resonate.

In stark contrast, IDEs have mastered the art of learning through iteration. Their approach is not about launching the perfect product but rather systematically uncovering what works, what doesn't, and why, long before scaling up. At the heart of this approach lies the *experimental innovation loop*, a repeating cycle of hypothesis formulation, structured experimentation, data collection, and reflective evaluation.

Each loop is a micro-journey in itself. It begins with a hypothesis—perhaps an assumption about customer pain points, user behavior, or technological feasibility. This is then tested through carefully designed experiments: prototypes, Minimum Viable Products (MVPs), A/B tests, or pilot rollouts. But experimentation alone isn't enough. What follows is a rigorous evaluation stage, where evidence is interpreted and used to either validate, refine, or discard the original hypothesis. Only when learning is complete does the next loop begin with greater clarity and deeper insight.

As illustrated in Figure 4.1, IDEs progress through these iterative loops while navigating two critical forces: uncertainty and resource commitment. At the inception phase, uncertainty is at its peak—both in terms of market understanding and operational constraints—yet the resources required to

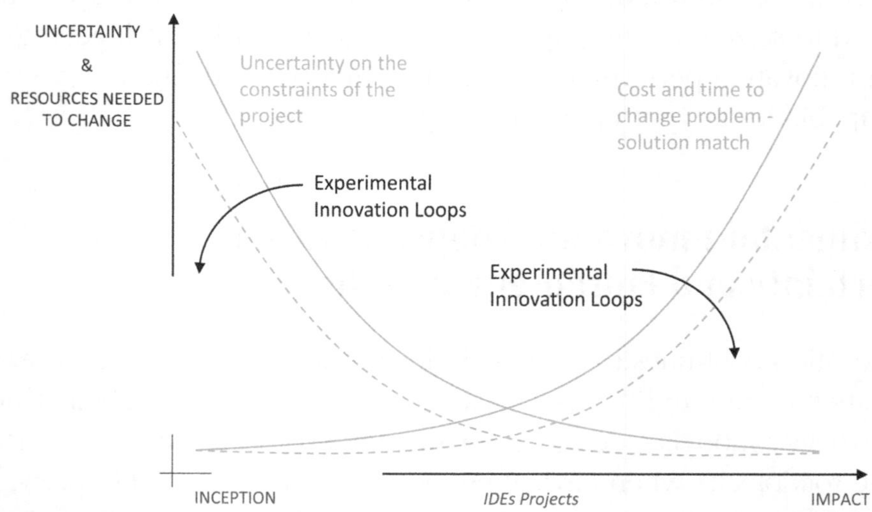

Figure 4.1 Experimental innovation loop.

pivot are relatively low. As projects move toward impact, this balance reverses: decisions become more expensive, and the cost of failure rises. Without early experimentation, any pivot at this stage can be prohibitively costly.

This is precisely why IDEs sequence their innovation journey in loops with each one strategically reducing uncertainty before increasing investment. Rather than attempting to control all variables upfront, they embrace ambiguity as a source of discovery. The impact of this model is not just technical; it reshapes the economics of innovation. Each validated loop serves as a milestone, a de-risking checkpoint that reassures investors, co-founders, and collaborators alike.

Over time, these loops enable a measured, evidence-based path to growth. They help IDEs transition from seed-stage ventures to Series A, B, and C funding rounds, attracting investors who are increasingly risk-averse yet hungry for scalable, validated ideas. From a financial standpoint, this milestone-driven approach converts uncertainty into traction, translating qualitative learning into quantitative performance indicators.

Importantly, this methodology is not limited to startups. Large corporations that engage with or learn from IDEs can embed similar experimental loops into their own innovation frameworks. Doing so allows them to stay nimble in volatile markets, respond to disruptive trends with agility, and partner more effectively with entrepreneurial ventures. In a world where the only constant is change, the capacity to learn fast—and act faster—is what separates organizations that thrive from those that fade.

In the end, experimental innovation loops represent more than just a technique; they embody a mindset shift: one where failure is not feared but used as feedback: one where progress is measured not just by output but by insight, and one where managing uncertainty becomes not a defensive act but a creative engine for growth.

Adaptability as a Response to Innovation Uncertainty

In the evolving landscape of innovation, uncertainty is not an anomaly; it is the norm. Whether it's a new technology still in flux, an emerging ecosystem with undefined roles, or an untested business model, uncertainty infiltrates every layer of disruptive innovation. Scholars have categorized this uncertainty into three principal domains: *technological*, *ecosystemic*, and *business model uncertainty*. Startups tend to navigate these turbulent waters with agility, driven by lean experimentation and iterative learning.

By contrast, established firms—even with their abundant capital and talent—often struggle. Bureaucratic inertia and resistance to change can erode responsiveness. For this reason, hybrid collaborative models have emerged, such as alliances between traditional automakers and autonomous vehicle startups, to co-manage these uncertainties in a shared-risk, co-learning environment (Kapoor & Klueter, 2020).

At the heart of effective uncertainty management lies adaptability. Innovation cannot thrive in static systems governed solely by traditional analysis and long-term forecasting. Instead, dynamic environments demand structured yet flexible decision-making frameworks. Organizations must employ *structured learning plans* to identify critical unknowns, ranging from technical feasibility and market validation to internal alignment and resource constraints. These plans serve as scaffolding for iterative testing, enabling learning cycles that refine assumptions and inform key decisions when it matters most (Rice et al., 2008).

However, tackling uncertainty isn't only a structural or strategic challenge; it is also deeply psychological. Managers, driven by personal accountability and career preservation, often favor incremental improvements over bolder, riskier bets. While such caution might ensure short-term stability, it tends to suppress transformative innovation. To counteract this behavioral bias, organizations must operate at the portfolio level, balancing safer, incremental projects with more speculative initiatives (see section "Leadership for Organizational Innovation" in Chapter 1). Moreover, project-level decisions should be viewed in the broader context of strategic goals. This calls for institutional mechanisms that promote realistic risk assessments, open conversations about uncertainty, and independent decision-making bodies that minimize bias and short-sightedness. The aim is to embed a culture of responsible experimentation—one that not only tolerates but also encourages calculated risk-taking (Lovallo et al., 2020).

In fast-changing markets, adaptability becomes indispensable. Adaptive systems are those capable of responding autonomously and continuously to changing environmental conditions. These systems, which evolve their behavior based on ongoing inputs, stand as the vanguard of innovation resilience. Yet, despite progress in machine learning and adaptive algorithms, achieving true autonomy—where systems adapt without human intervention—remains a frontier of research.

A useful conceptual framework here distinguishes between *"known unknowns"* and *"unknown unknowns."* The former includes challenges that the system hasn't encountered but can respond to using past analogs and existing reasoning structures. The latter represents a more profound difficulty: unanticipated scenarios that lie outside the current experience base. Addressing these requires real-time adaptive strategies and cognitive agility.

Key to this evolution is the development of *autonomous environmental representations*. Systems must independently identify which variables in the environment are significant and worthy of response. Although early research has made strides through enhanced state-space representations, the challenge persists: reliably discerning which elements are noise and which carry meaning. Lifelong learning further amplifies the complexity. Self-adaptive systems must not only store and retrieve past knowledge but also accelerate their learning in new contexts using tools such as *transfer learning* and *behavioral repositories*. These are not mere technical capabilities—they are the backbone of resilience in autonomous systems.

Beyond responsiveness, adaptive systems must be capable of redefining their goals. When original objectives become obsolete due to unforeseen shifts, the system must evaluate the feasibility of continuing, revising, or entirely redefining its targets. This ability to reanchor intent is critical to sustaining relevance in unpredictable environments. However, this introduces profound ethical implications. If systems are allowed to autonomously reshape their goals, who ensures these changes align with human values and safety protocols? Innovation management systems must therefore prioritize not only adaptive functionality but also the ethical architectures that govern it (Dusparic & Cardozo, 2021).

In an organizational setting, adaptability isn't about technology alone—it's about mindset, structure, and culture. Adaptive organizations accept uncertainty as a constant, not an exception. They redesign their operations, leadership approaches, and resource flows to flex in response to both predictable shifts and abrupt disruptions (Reeves & Deimler, 2011). Yet many large enterprises remain locked in rigid models, optimized for control and efficiency. These models work well in stable markets but falter under conditions of ambiguity, change, and disruption.

To bridge this gap, companies must embrace *organizational ambidexterity*—the capacity to pursue both incremental (Small-I) and disruptive (Big-I) innovation simultaneously (Kleiner, 2013). Most large organizations are competent at Small-I: they excel in incremental improvements, guided by stage-gate processes and front-loaded risk assessments. But Big-I innovation—where solutions and business models are in flux—requires looser, boundary-crossing structures. Some firms have risen to this challenge. For instance, Noblis employs *internal crowdsourcing* to surface unconventional ideas, while General Motors fosters emergent innovation through *inter-organizational events* that create adaptive spaces for ideas and resources to flow freely.

Adaptability must also extend to performance measurement (see section "Measuring Innovation Performance: A Balanced, Strategic, and Adaptive Approach" in Chapter 5). Traditional KPIs fall short in high-uncertainty contexts. What's needed is a multidimensional system that evaluates innovation across the full lifecycle. This includes inputs (resources, skills), throughputs (experimentation rates, collaboration efficiency), outputs (time-to-market, solution readiness), and outcomes (financial returns, stakeholder impact, social value). Critically, such metrics must capture not only short-term deliverables but also medium- and long-term shifts. The *Balanced Scorecard* provides one such framework, offering integrated views across financial, customer, process, and learning dimensions to align operational innovation with broader strategic goals (Merton, 2013).

At the core of this adaptive capacity lies the development of *dynamic capabilities*—the organizational muscle to integrate, build, and reconfigure internal and external competencies in response to changing conditions (Teece, cited in Kleiner, 2013). These are not static strengths but fluid capabilities, shaped and reshaped through learning, experimentation, and feedback. Leadership plays a pivotal role. Adaptive leaders do more than approve plans—they cultivate cultures that embrace ambiguity, reward initiative, and balance exploratory ambition with operational execution (see section "Future-Focused Leadership" in Chapter 1).

Ultimately, full adaptability is not optional. It is the bedrock of resilience in an era defined by uncertainty. Achieving it demands not only advanced technical systems that can respond to the unknown but also organizations brave enough to relinquish outdated assumptions and invest in continuous reinvention. It is this fusion—of autonomous innovation systems and human-led, culture-driven adaptability—that transforms uncertainty from a constraint into a strategic frontier.

Building Organizational Adaptability to Manage Innovation Uncertainty

We live in an age where rapid technological evolution and market volatility are the new constants. To thrive in this context, organizations must move decisively beyond efficiency-centric paradigms and embed adaptability at the core of their strategic DNA. Innovation uncertainty is no longer a sporadic challenge; it's a systemic condition, driven by unknown technologies, shifting customer expectations, and fluid ecosystems. Navigating this reality demands structures and processes that are not only resilient, but also capable of evolving with speed and intention.

This section explores how large organizations can cultivate such adaptability. The transformation begins with three key levers: *adaptive space, ambidextrous organizational design*, and a *dual innovation system* that balances exploitative and exploratory approaches. These mechanisms serve as the scaffolding for transitioning from traditional, efficiency-driven innovation systems—known as Small-I—toward more agile, entrepreneurial modes—Big-I. When strategically integrated, they allow companies to build sustained innovation capabilities even in the face of persistent uncertainty.

Most established firms are historically built to support Small-I innovation. This means they favor incremental improvements, emphasize cost efficiency, and follow well-defined operational routines. These models may excel at sustaining core business performance but often fall short when the goal is radical innovation or disruptive growth. For a company to shift toward a Big-I orientation, it must first acknowledge the structural and cultural differences between these two innovation systems.

Research shows that organic, innovation-led growth tends to be more profitable and sustainable than growth achieved through acquisition. However, for many large firms, unlocking that emergent innovation from within requires deliberate, even radical, change. While tech giants like Alphabet, Apple, and Meta are often heralded for their fluid, innovation-friendly environments, many firms in other industries remain burdened by rigid hierarchies that inadvertently stifle innovation from the inside out.

At the heart of Big-I innovation lies the concept of *adaptive space*—a dynamic organizational context that enables people, ideas, resources, and information to flow freely across boundaries. This space serves as a vital connective tissue, linking the entrepreneurial "edges" of the organization with its operational core. It is where new ideas are incubated, shaped, and amplified until they are ready to be absorbed into the broader operating system.

Adaptive space should not be seen as a fixed physical place. Instead, it is a networked environment—a permeable layer within the organization that flexes and shifts according to evolving innovation needs. Some companies create this space through internal crowdsourcing platforms, as the American not-for-profit corporation, Noblis has done. Others, like General Motors, build it through cross-functional innovation events that bring diverse perspectives together. In both cases, the goal is the same: enable emergent ideas to travel, grow, and connect with the operational engine that can bring them to market.

To support this, organizations must adopt *ambidextrous structures*—ones that allow exploitative (Small-I) and exploratory (Big-I) systems to coexist. This doesn't simply mean adding a new unit or innovation lab. It means designing distinct, differentiated systems with their own rules, processes, and leadership styles, while maintaining strategic alignment at the top. Senior leaders play a pivotal role in ensuring these parallel tracks remain connected and coherent.

Figure 4.2 illustrates *dual innovation system* as the contrasts between Small-I and Big-I systems across ten dimensions. While Small-I innovation operates in structured, hierarchical environments focused on incremental gains and cost efficiency, Big-I thrives in fluid, adaptive contexts. Its goals are breakthrough and emergent innovation, driven by teams that are loose, entrepreneurial, and oriented toward long-term growth. Big-I systems prioritize learning over control, experimentation over predictability, and risk-taking over risk aversion. In doing so, they offer organizations a pathway to sustained relevance in uncertain environments.

	Small-I – Business-as-Usual	Big-I – New Emerging Business
Focus	Exploitative business	Exploitative business *(engaging with IDEs, UNI Lab-based startups, STEM Talents)*
Organizational Context	Focus on internal channels of communication and authority	"Adaptive space", fluid
Strategic Goals	Cost/profit	Innovation/growth
Innovation Goals	Sustaining, incremental innovation	Breakthrough, organic, emergent innovation
Innovation Projects	Stage-gate approach	Innovation-loop approach (I-Cap)
Critical Tasks	Operations efficiency	Result-achievement efficacy
Competencies	Operational/managerial	Entrepreneurial/innovation-driven (E-Cap)
Structure	Formal/hierarchical	Loose/flat
Rewards	Productivity	New ideas
Culture	Low risk	Risk-taking

Figure 4.2 Small-I vs. Big-I innovation system. (From O'Reily III, C. A., & Tushman, M. L., 1996. With Permission.)

The coexistence of these systems—when managed well—offers the best of both worlds. But managing innovation under this duality is no small feat, especially when it comes to large-scale projects. Many established firms still rely heavily on stage-gate processes, which emphasize rigorous upfront planning. In these models, vast resources are allocated early in the innovation journey, often before sufficient validation of the underlying assumptions. Risks are assessed and mitigated in a linear sequence: first customer research, then technology feasibility, followed by full-scale development. But such an approach can backfire, especially in Big-I scenarios where uncertainty is highest precisely in the early stages.

By contrast, IDEs—explored in depth in Chapter 5, section "Managing Disruptive Technologies: Contrasting Approaches of Large Enterprises and Innovation-Driven Enterprises (IDEs)"—flip this script. They minimize early capital exposure, prioritize rapid experimentation, and iterate in short, evidence-based cycles. Rather than making large bets upfront, they build conviction incrementally through validated learning. Their funding strategies reflect this logic as well: raising capital in rounds aligned to clear innovation milestones.

For large organizations looking to adopt similar practices, the shift is as much cultural as it is structural. It requires moving away from a mindset of predict-and-plan to one of sense-and-respond. This means creating space for uncertainty, investing in learning, and recognizing that adaptability is not a one-time effort, but an ongoing capability.

In the end, building adaptability is not just about keeping up with change; it's about staying ahead of it. It's the mechanism through which uncertainty becomes not a threat, but a driver of discovery, reinvention, and long-term advantage.

Business Model Innovation as an Adaptive Strategy

In a world where change is the only constant, business model innovation has emerged not merely as a competitive advantage but as a critical strategic response to uncertainty. As markets grow more dynamic and disruptions more frequent—whether technological, behavioral, or geopolitical—organizations must be equipped to rethink the very architecture of how they create, deliver, and capture value.

Business model innovation isn't just a buzzword or a matter of tweaking a few operational levers. It is a fundamental reconfiguration of the enterprise logic.

At the heart of this rethinking lies the classic framework introduced by Peter Drucker, whose conceptualization of the business model remains a touchstone in the field. According to Drucker, five core elements constitute the backbone of a business model: the customer, the value proposition, the offering (product or service), the value creation mechanism, and the value appropriation mechanism (Drucker, 2012 [1954]). Together, these components dictate how an organization functions, competes, and evolves (see also section "Business Model Innovation and Value Creation" in Chapter 2).

Customer centricity sits at the core of Drucker's philosophy. It all begins with understanding not just what customers want but how they define value and experience benefit. Solving a pressing need or offering a unique advantage often lies at the center of any compelling value proposition. The value creation mechanism, in turn, translates that promise into operational reality. This means orchestrating the right mix of internal capabilities, external partnerships, and efficient workflows to ensure sustainable delivery. But creating value is only half the equation; capturing it is equally vital. Organizations must generate financial returns sufficient to support reinvestment, scale, and long-term viability.

Drucker identified two main dimensions of value capture. First, *revenue mechanisms*: these dictate how, when, and through what channels customers pay. Here, strategic pricing models, timing of payment, and bundling strategies become essential levers. Second, *protection mechanisms*: these serve to safeguard intellectual property and prevent competitors from easily imitating the innovation. This is the terrain of appropriability regimes—how well a firm can defend and sustain its competitive advantage (see "Market Dynamics and Strategic Trade-Offs" in Chapter 2).

Business model innovation can take two forms. It might involve incremental refinements—adjusting individual elements while maintaining the overall architecture. Or it might manifest as a radical reinvention, disrupting entire industries by reshaping the relationships among the elements of value. While the former is more common and less risky, the latter, when executed well, holds the potential for massive strategic gain. Such transformations can redefine how value is produced, captured, and perceived by both customers and competitors (Teece, 2010).

To further understand this dynamic, it is helpful to consider the three archetypes of value configuration: the *value chain*, the *value shop*, and the *value network* (Stabell & Fjeldstad, 1998). The value chain, common in manufacturing, involves a linear progression from input to output. The value shop, more typical of consulting or healthcare, revolves around diagnosing and solving problems through expert knowledge. The value network is built on interconnection, linking customers, partners, and resources in ways that generate value through participation and exchange. It is foundational to digital platforms, service marketplaces, and peer-to-peer ecosystems.

Take, for instance, Apple's iPod and its strategic pairing with the iTunes Music Store. The device alone—offering "1,000 songs in your pocket"—addressed a clear consumer desire. But the real innovation came in 2003 when Apple introduced the iTunes platform. It wasn't just a product; it was a two-sided ecosystem integrating hardware, software, and licensed content. The pricing model—$0.99 per song—disrupted the traditional music industry and made digital music accessible, legal, and convenient. Seamless synchronization further elevated the user experience. From a business model perspective, Apple moved beyond one-off device sales to recurring content revenue. On the protection side, Digital Rights Management (DRM) and user account linkage created significant switching costs. Within three years, this integrated system evolved into a $10 billion business, reshaping not only Apple's balance sheet but also the entire music industry.

By contrast, Google Glass offers a cautionary tale about the pitfalls of misaligned business models. Although the technology was groundbreaking, offering voice-controlled, head-mounted displays, the go-to-market strategy failed to match the innovation. Priced at $1,500 and lacking a supporting app ecosystem, Glass offered little incentive for mass adoption. There was no graduated pricing, no subscription service, no try-before-you-buy model. Moreover, privacy concerns and social discomfort further deterred consumers. In short, Google focused on the product but neglected the platform. Despite its technological edge, Glass faltered because its business model failed to resonate with user expectations and economic logic.

In the digital era, *data-driven innovation* (DDI) has emerged as a powerful enabler of business model transformation. DDI refers to the strategic use of large, diverse data sets to generate economic and social value through smarter decision-making, operational efficiency, and entirely new modes of value creation (OECD, 2015). The proliferation of digital platforms, Internet of Things devices, and cloud computing has unlocked vast reservoirs of actionable data. Increasingly, companies are using this data not only to optimize operations but also to reinvent their business models from the ground up.

Duolingo offers a compelling example of how data and business models can converge. Through gamification and adaptive learning algorithms, the language-learning platform personalizes instruction based on user proficiency, pace, and engagement levels. This personalization drives retention and enhances learning outcomes. More impressively, Duolingo diversified its revenue model: beyond free usage, it monetized via ads, subscription tiers, and a certified language proficiency test—a service that has grown into a major revenue stream (Duolingo Inc., 2022).

Yet while DDI opens doors, it also presents formidable challenges. It demands specialized talent—data scientists, AI engineers, and cybersecurity experts—who are in short supply and high demand. For Duolingo, maintaining trust through responsible data practices is just as important as refining its learning algorithms. Technological readiness, organizational alignment, and cultural openness all play key roles in translating data into sustainable business value.

When it comes to value capture, data-driven business models operate on three primary fronts. First, by enhancing existing revenue streams through efficiency gains or increased customer lifetime value. Second, by deepening customer insights and enabling refined product offers or upsell opportunities. Third, by uncovering entirely new markets, applications, or customer segments. These mechanisms often evolve in tandem as an organization's data capabilities mature.

Data-driven strategies are also gaining traction in manufacturing, particularly in the context of cyber-physical systems and human-centric innovation. Firms are now leveraging real-time data to enable mass customization, predictive maintenance, and dynamic supply chain management. These innovations demand corresponding shifts in business models. Traditional production-based revenue structures no longer capture the full scope of value. Instead, manufacturers are exploring service-based, outcome-oriented models that align better with their evolving capabilities and customer expectations (European Commission, 2021).

However, business model innovation rarely happens in a vacuum. It often requires a reprogramming of the organization's very DNA from culture and incentives to resource allocation and governance structures. Companies must cultivate ecosystems where value is co-created with customers, partners, and even competitors. They must continuously reevaluate their business logic in light of new threats, emerging technologies, and shifting social norms.

In short, business model innovation is not an occasional strategic pivot; it is a continuous act of organizational evolution. Whether through incremental refinements or radical reinvention, companies that master this discipline will be best positioned to thrive in the face of uncertainty. As industry boundaries blur and digital technologies reshape every sector, the capacity to reimagine how value is created and captured will remain a defining feature of sustainable success.

Financing Innovation: Approaches and Models

Capital Markets and Collaboration Dynamics Between IDEs and Risk Capital Providers

Excellence in innovation management does not stem from creativity alone. It requires a strategic approach to financing—one that can absorb the uncertainty intrinsic to innovation while remaining adaptive to fast-moving developments. Innovation, after all, does not unfold in a vacuum; it must be funded, staged, and supported by a broader financial ecosystem capable of navigating risk and unlocking value.

Capital markets play a vital role in this equation. They are not just arenas for profit-seeking speculation, but serve as key conduits that channel financial resources from households and institutional investors into entrepreneurial ventures and breakthrough technologies. Understanding how capital markets operate is essential to grasping how innovation gets financed, how risk is distributed, and how resource flows are aligned with the maturity and potential of emerging ideas.

At the foundation of these markets, financial resources are supplied by households—often indirectly—through pension funds, retirement accounts, and personal investment vehicles. These funds are aggregated and managed by institutional investors, which make up the "buy side" of the capital markets. These include pension funds, mutual funds, hedge funds, sovereign wealth funds, and university endowments. Although united by their role in capital allocation, they differ significantly in terms of risk appetite, time horizons, and regulatory frameworks. Mutual funds, for instance, typically seek stability through diversification, while hedge funds may pursue aggressive, high-risk strategies with greater use of leverage and alternative assets.

Working in tandem with them is the "sell side"—the domain of investment banks, broker-dealers, and financial analysts. These entities structure deals, facilitate trading, underwrite public offerings, and provide strategic advisory to both investors and entrepreneurs. Analysts play a pivotal role by producing insights on industries and companies that guide investment flows. Meanwhile, financial media and public discourse shape perceptions and expectations subtly but powerfully influencing valuation and investor sentiment.

For established, publicly listed companies, capital markets become the real engine of innovation because the quarterly share price is largely shaped by investors' expectations of future earnings from innovation projects that these firms deliberately signal to the "buy side" in order to demonstrate credible innovation capability, stimulate demand for the stock, and ultimately sustain or increase market cap.

It is within this intricate ecosystem that IDEs must navigate, operating under a fundamentally different financial logic than established firms. Whereas, listed incumbents can leverage public market expectations and quarterly signaling to sustain innovation investment, IDEs typically build credibility from the ground up, progressing without the benefit of an established revenue base, stable cash flows, or a mature market narrative. Unlike traditional firms that follow linear development models, IDEs adopt an iterative, hypothesis-driven approach. Innovation unfolds in cycles—loops of experimentation, learning, validation, and pivoting. This dynamic and adaptive methodology, as explored earlier in this chapter, demands financial models that are equally flexible. The capital must arrive in waves, increasing in volume and commitment only as the evidence of market fit and feasibility becomes more robust.

Herein lies the essence of *venture capital* (VC) and early-stage financing. These forms of risk capital are designed to mirror the cadence of innovation. At the early stages—often called pre-seed or seed funding—investors provide relatively small amounts of capital. These resources are used to test assumptions, build MVPs, and validate early demand. This is followed by Series A, where the focus shifts toward scaling initial traction and refining the business model.

As IDEs demonstrate increasing market acceptance and operational maturity, they move into later-stage financing rounds (Series B, C, D, etc.). The stakes are higher, not just in terms of capital, but also in the expectations for growth, speed, and return on investment. These rounds often attract larger venture firms or strategic investors who bring not only funding but also expertise, networks, and governance support. The financial commitment intensifies as the innovation moves closer to commercialization and competitive entry. Strategic alignment between investors and IDEs becomes paramount.

Venture capitalists are not merely financiers; they are often co-creators. Their engagement includes mentorship, operational guidance, recruitment support, and connections to key partners or customers. This active involvement mitigates entrepreneurial risk and increases the odds of scaling innovative ideas into market-ready solutions. Trust, transparency, and a shared understanding of uncertainty are vital ingredients of this partnership.

Adaptability is essential for both sides. IDEs must remain agile enough to integrate market feedback and investor insights without losing their creative edge. Investors, in turn, must fine-tune their strategies to accommodate evolving business models, shifting timelines, and technological volatility. Innovation financing is not a static transaction; it is a collaborative, living relationship shaped by evidence, experimentation, and evolving context.

A Hypothetical Example: Dilution, Growth, and Strategic Trade-offs

To bring this dynamic into sharper focus, consider a simplified example of startup financing through three stages: Initial, Series A, and Series B, as illustrated in Figure 4.3.

At inception, a startup is self-funded by its founders, who contribute personal capital. The balance sheet shows $100 in cash and $300 in operating assets. At this point, there are 100 shares issued, each implicitly valued at $4, reflecting a total founder equity of $400 and 100% ownership. This valuation, however, is only theoretical. Without an external transaction, there is no market-based price for the company's equity.

In the Series A round, the startup seeks $100 in new capital to support a validated growth opportunity. A venture investor agrees to provide this amount in exchange for a 20% stake. To facilitate this, the company issues 25 new shares, increasing the total to 125. The company is now valued at $500 post-money. The founders retain 80% of the ownership, while

Stage	Assets	Liabilities & Net Worth	Shares & Values
Initial (Pre-Funding)	Cash: $100 Operating Assets: $300	Equity (Founders): $ 400	100 shares, $4/share
Series A Funding	Cash: $200 Operating Assets: $300	Equity (Founders): $ 800 Equity (Series A Investors): $100	125 shares, $4/share
Series B Funding	Cash: $1,000 Operating Assets: $1,000	Equity (Founders): $ 400 Equity (Series A Investors): $200 Equity (Series A Investors): $1,000	250 shares, $8/share

Figure 4.3 Venture capital financing stages: Initial, Series A, and Series B.

the Series A investor holds 20%. Crucially, the valuation becomes explicit, anchored by the investor's willingness to pay.

In the Series B round, the startup pursues a much larger capital injection—$1,000—to scale operations significantly. New investors provide this funding in exchange for a 50% stake, implying a $2,000 post-money valuation. To reflect this new structure, 125 additional shares are issued, bringing the total to 250 shares. The new share price rises to $8. At this point, the founder's ownership is reduced to 20%, but its market value has doubled from $400 post-Series A to $800 post-Series B. The value of their equity has increased even as their control has diluted.

This staged progression illustrates the fundamental logic of venture financing: ownership dilution is a trade-off for accelerated growth and value creation. The founders give up control but, in return, gain access to resources, support, and scale that would otherwise be out of reach. This model works because each stage builds on validated milestones, and each new investor enters at a higher valuation, driven by evidence of traction and reduced uncertainty.

In sum, financing innovation is not just a matter of funding projects. It is about designing capital strategies that are sensitive to uncertainty, aligned with learning cycles, and capable of evolving in lockstep with the innovation journey. The collaboration between IDEs and their capital providers is at the heart of this process, one that, when managed wisely, transforms early ideas into market-defining ventures.

Beyond Budgeting: Adaptive Financial Models for Innovation in a Dynamic Environment

In environments where uncertainty is the norm and change is constant, adaptability becomes a critical organizational competency. One of the areas where this need for agility is most urgent—and most often constrained—is in financial planning. Innovation does not unfold on a predictable timeline, nor does it conform to the rigid frameworks of traditional budgeting processes. Fixed annual plans, predetermined resource allocations, and inflexible performance targets can become obstacles rather than enablers of innovation.

To support the exploratory, nonlinear nature of innovation, organizations must evolve their financial models. They must shift from static control systems to dynamic resource orchestration where funds are mobilized not according to fixed budgets but in response to real-time learning, emerging priorities, and validated progress. Adaptive financial planning involves a blend of flexible operational planning, agile decision-making, and budgeting processes that mirror the rhythm of innovation cycles.

Traditional budgeting tends to reinforce hierarchical decision-making and fixed cost structures. These models may work well in stable environments, but they fall short when innovation is the strategic priority. In contrast, alternative financing models—such as venture funding and milestone-based resource allocation—offer more flexible, responsive approaches. These models are particularly well-suited to situations of high unpredictability, allowing organizations to invest incrementally as innovation initiatives mature and generate evidence of potential value. As Gompers et al. (2021) point out, the venture funding model provides not only capital but also a governance structure based on learning and adaptation.

This is especially true for IDEs, which operate in iterative, hypothesis-driven cycles. Their funding needs are not linear but evolve in response to experimental feedback and market signals. IDEs require financial mechanisms that are structured enough to provide discipline but flexible enough to accommodate pivots, delays, and discoveries. One effective practice emerging from entrepreneurial ecosystems is phased financing—the allocation of funds tied to specific learning milestones or development stages.

Under this approach, investors or funding bodies provide capital in stages. Early-stage investors, such as those participating in seed or Series A rounds, typically contribute smaller amounts to support early testing and initial market exploration. Later-stage investors step in once the IDE has demonstrated traction, providing more substantial resources to fuel growth, scale operations, or expand into new markets. The entire financing journey is calibrated to the venture's evolving maturity, with each stage de-risking the next. This staged alignment between capital and innovation cycles is also applicable to internal contexts.

Indeed, organizations themselves can emulate venture funding principles through internal innovation financing structures. One increasingly popular model for established companies is the creation of an *internal venture capital fund*, a dedicated pool of financial resources earmarked for high-potential exploratory projects. These internal funds can be set up as a percentage of company revenue, profit, or even tied to strategic reinvestment goals. What makes them powerful is not just the money they provide but the agility they enable. Funds are released in tranches, aligned with predefined milestones, and can be reallocated as learning unfolds. In this way, organizations can nurture innovation without locking resources into static project plans, and without waiting for annual budgeting cycles to catch up.

This model also strengthens the internal innovation culture. It signals that innovation is not a side activity but a core strategic function deserving of dedicated resources and disciplined experimentation. It builds internal governance mechanisms that mirror the best of VC: performance-driven, evidence-based, and focused on long-term value creation.

Beyond internal funding, *crowdfunding* has emerged as a compelling alternative for financing innovation—one that not only raises capital but also engages the market directly. Unlike structured VC, crowdfunding draws on voluntary contributions from external stakeholders, including early adopters, customers, and interested members of the public. These contributors may or may not expect returns, and if they do, the compensation varies widely: equity, product discounts, exclusive access, revenue shares, or simply early visibility into a compelling innovation.

Though commonly associated with startups, crowdfunding also holds great potential for established companies. When used strategically, it enables organizations to simultaneously pursue financial and strategic objectives: raising funds, validating market demand, testing product hypotheses, and generating pre-launch buzz. Importantly, it offers early consumer feedback, helping to assess real-world interest in an idea before full-scale investment.

For larger firms, some reward mechanisms used in crowdfunding—such as equity distribution—may pose legal or structural barriers. However, non-equity rewards like early access, limited editions, or co-creation opportunities remain accessible and valuable. In this sense, crowdfunding serves as more than a fundraising tool. It becomes a market-sensing instrument and a participatory platform that expands the innovation ecosystem beyond organizational boundaries.

When combined, internal venture funds and external crowdfunding mechanisms offer a powerful dual approach: one grounded in strategic alignment and the other in public engagement. Together, they diversify the innovation portfolio, de-risk investments, and help organizations build more resilient and inclusive innovation systems (see section "Leadership for Organizational Innovation" in Chapter 1).

To support these evolving practices, many organizations are also adopting the principles of the *Beyond Budgeting* movement. This approach replaces static annual budgets with rolling forecasts, decentralized decision-making, and performance evaluations based on relative and contextual targets. Rather than fixing resource allocations in advance, organizations monitor real-time developments and reallocate funds dynamically, in response to emerging data and opportunities (Stange et al., 2021). Companies such as Roche have implemented Beyond Budgeting principles to enhance agility, responsiveness, and innovation capacity. The approach enables faster decision-making, better alignment with strategic priorities, and increased trust in frontline managers to act on local knowledge.

Moreover, platforms that integrate crowdfunding into innovation strategies also enable ongoing market experimentation. As Cornelius and Gokpinar (2021) highlight, crowdfunding campaigns offer opportunities to test market signals, adjust product features, and build communities of committed early adopters—all before a product formally launches. This shortens feedback loops and provides real-world validation that goes beyond internal assumptions.

In conclusion, moving beyond traditional budgeting is not just a financial reform; it is a strategic shift. It redefines how resources are deployed in support of innovation. Through phased financing, internal venture models, and crowdfunding, organizations can embrace financial agility while staying aligned with long-term innovation goals. Combined with the Beyond Budgeting philosophy, these models empower enterprises to adapt faster, scale smarter, and create more meaningful value—even amid the uncertainty that defines today's competitive landscape.

References

Cornelius, P., & Gokpinar, B. (2021). Crowdfunding Can Deliver More Than Just Money. *Harvard Business Review.*

Drucker, P. (2012). The Practice of Management. *Routledge* (first published in 1954).

Duolingo Inc. (2022). Form 10-K: Annual Report for the Fiscal Year Ended December 31, 2022.

Dusparic, I., & Cardozo, N. (2021). Adaptation to Unknown Situations as the Holy Grail of Learning-Based Self-Adaptive Systems: Research Directions. *SEAMS 21, ACM.*

European Commission (2021). Industry 5.0: Towards a Sustainable, Human-Centric and Resilient European Industry. *Directorate-General for Research and Innovation, Publications Office of the European Union.*

Gompers, P., Gornall, W., Kaplan, S. N., & Strebulaev, I. A. (2021). How Venture Capitalists Make Decisions. *Harvard Business Review.*

ISO (2019). ISO 56002:2019 – Innovation management system – Guidance. *International Organization for Standardization.*
ISO (2020). ISO 56000:2020 – Innovation management – Fundamentals and vocabulary. *International Organization for Standardization.*
ISO (2024). ISO 56001:2024 – Innovation management system – Requirements. *International Organization for Standardization.*
Kapoor, R., & Klueter, T. (2020). Innovation's Uncertainty Factor. *MIT Sloan Management Review.*
Kleiner, A. (2013). David Teece's Dynamic Capabilities. *Strategy+Business.*
Lovallo, D., Koller, T., Uhlaner, R., & Kahneman, D. (2020). Your Company Is Too Risk-Averse. *Harvard Business Review.*
Merton, R. C. (2013). Innovation Risk: How to Make Smarter Decisions. *Harvard Business Review.*
OECD (2015). Data-Driven Innovation: Big Data for Growth and Well-Being. *OECD Publishing.*
Reeves, M., & Deimler, M. (2011). Adaptability: The New Competitive Advantage. *Harvard Business Review.*
Rice, M. P., O'Connor, G. C., & Pierantozzi, R. (2008). Implementing a Learning Plan to Counter Project Uncertainty. *MIT Sloan Management Review.*
Stabell, C. B., & Fjeldstad, Ø. D. (1998). Configuring value for competitive advantage: on chains, shops, and networks. *Strategic Management Journal, 19(5)*, 413–437.
Stange, S., Bogsnes, B., & Sheth, H. (2021). The Future is Beyond Budgeting. *Boston Consulting Group.*
Teece, D. J. (2010). Business models, business strategy and innovation. *Long Range Planning, 43(2–3)*, 172–194.

Chapter 5

A Systems Approach to Innovation Management

The Systems Approach According to ISO 56000 Series Requirements

Innovation demands intentional design, coherent structures, and coordinated practices that evolve over time. To navigate complexity and foster sustained innovation, organizations must move beyond ad hoc initiatives and adopt a systems approach—a structured, interconnected way of managing innovation as a dynamic, adaptive ecosystem. This is the foundational logic behind the International Organization for Standardization (ISO) *56000 series*, which articulates a comprehensive framework for innovation management systems (IMS).

At its core, the systems approach emphasizes that innovation is not a standalone event or a creative burst confined to research and development (R&D) labs. Rather, it is a collective organizational capability—an ongoing interplay of processes, people, knowledge, resources, stakeholders, and outcomes, each influencing and reinforcing the others. A systems perspective enables organizations to manage these elements holistically, uncover interdependencies, and strengthen coordination across silos.

According to *ISO 56001* (ISO, 2024), effective innovation management begins with strategic alignment and operational clarity. In *Clause 8*, the standard requires organizations to clearly define the scope, objectives, roles, responsibilities, decision-making structures, and required resources for each innovation initiative. This level of clarity creates a shared purpose while

DOI: 10.4324/9781003743583-5

enabling operational flexibility—both of which are essential when dealing with the inherent uncertainty of innovation work.

Moreover, systematic planning under the *ISO 56000* (ISO, 2020) *series* involves more than structuring efforts. It compels organizations to embrace adaptive governance. This includes processes for managing uncertainty, capturing critical knowledge, and responding to emerging challenges. Adaptive management helps organizations remain open to new insights and shifting conditions—a prerequisite for innovation in today's volatile environments.

Importantly, the systems approach does not end with execution. Continuous performance evaluation is a cornerstone of the framework, ensuring that innovation efforts are not only monitored but also continuously optimized. As outlined in *Clause 9* of *ISO 56001*, organizations are expected to define what they intend to measure and to select balanced innovation indicators across four key dimensions: inputs—such as the volume of ideas generated, time and talent invested, or budget allocated to innovation; productivity—including learning speed, iterative development efficiency, and the effectiveness of collaboration; outputs—for instance, the number of solutions tested, prototypes developed, or innovations brought to market; and outcomes—such as market impact, customer satisfaction, stakeholder value, and revenue derived from innovation.

This balanced scorecard approach ensures that short-term achievements do not overshadow long-term value creation and that operational metrics are linked to strategic intent. Performance monitoring is supported by internal audits, management reviews, and feedback loops that help keep innovation activities calibrated with external trends and internal priorities.

Yet, measurement alone is insufficient. The true power of the systems approach lies in how organizations use performance data to enable continuous improvement. *Clause 10* of *ISO 56001* addresses this by requiring that deviations, anomalies, and non-conformities be identified, investigated, and resolved through systematic corrective actions. When outcomes diverge from expectations, organizations must not only fix the immediate issue but also go deeper into analyzing root causes, learning from experience, and adapting practices to prevent recurrence.

This emphasis on learning—from both success and failure—is one of the most distinctive features of the systems approach. The documentation and dissemination of corrective actions play a critical role here. When shared transparently across the organization, these insights foster a culture of reflection and knowledge-based adaptation. Over time, this nurtures an innovation culture grounded not just in bold ideas but in the capability to learn, evolve, and execute reliably under changing conditions.

In sum, a robust IMS based on the *ISO 56000 series* integrates operations, performance evaluation, and continuous improvement into a coherent, living structure. By regularly reviewing progress through balanced indicators, adjusting plans in response to learning, and embedding feedback loops into the organizational fabric, companies position themselves for sustained innovation success. They become not only more innovative but also more resilient, coordinated, and future-ready—equipped to thrive in an environment where adaptability is the most valuable form of strategic advantage.

Innovation Systems: Structured Frameworks and Integrated Processes

Navigating the Evolution of Technology Through S-Curves and Readiness Frameworks

Technological innovation does not evolve in a straight line. More often, it follows a wave-like motion—an initial period of struggle and learning, followed by rapid breakthroughs, and eventually a slowdown as limits are approached. Organizations today find themselves increasingly challenged by the need to shift from incremental improvements of their current technologies to entirely new, disruptive paths that reset the rules of the game.

This progression is visualized through what's known as the *S-curve* of technology performance. As shown in Figure 5.1, a new technology typically begins its life with modest, often frustrating gains. In these early stages, understanding is shallow, experimentation is costly, and uncertainty runs high. This is the realm of incremental innovation—slow, deliberate refinements within a known technological space. As familiarity grows, however, development accelerates. Breakthroughs come more frequently, performance improves rapidly, and organizations harvest increasing returns on their investment in R&D. But this steep rise doesn't last forever. Eventually, the curve flattens as the technology nears its performance limits. Improvements become harder to achieve, costlier, and less impactful.

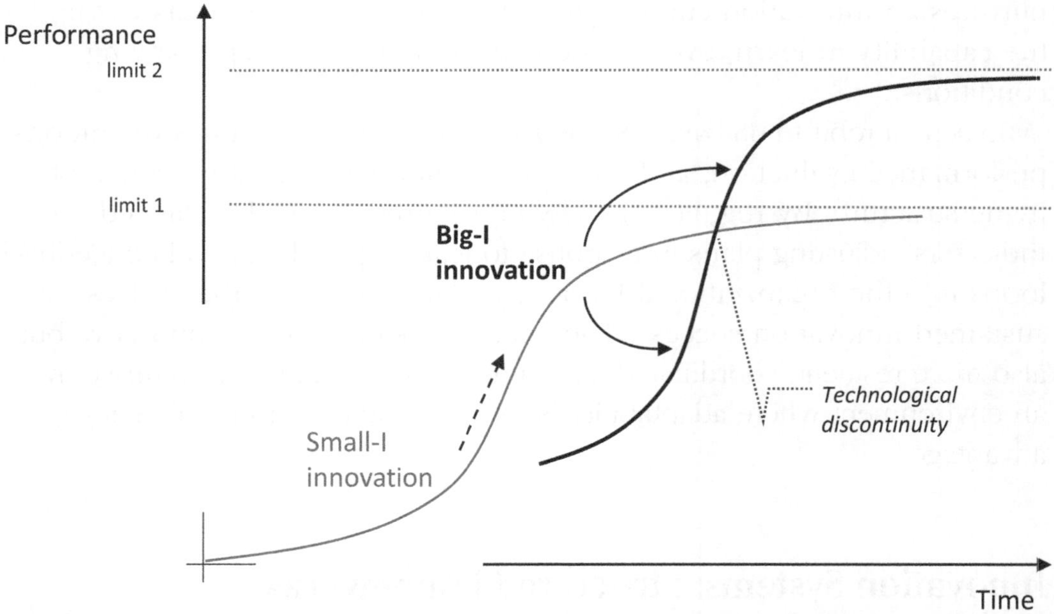

Figure 5.1 Technology progress S-curves.

What follows is a pivotal moment. Some organizations double down on the existing path, trying to wring out the last few drops of improvement. Others recognize the signs of diminishing returns and begin to explore alternatives—technologies that may at first seem immature or unproven but hold the promise of a new performance frontier. This is the realm of breakthrough innovation, where new paradigms are born. The point at which one S-curve gives way to another is known as a *technological discontinuity*. It's not only a moment of disruption, uncertainty, and risk but also of possibility.

The move from combustion engines to electric drivetrains in the automotive industry is a prime example. Traditional internal combustion engines, perfected over a century, have approached their engineering limits in terms of efficiency and emissions. Electric vehicles (EVs), initially criticized for short range and high costs, have over the last decade entered a rapid growth phase, supported by advancements in battery technology and global climate goals. Tesla, for instance, has successfully navigated this S-curve jump, transforming the landscape of mobility and forcing legacy automakers to follow suit.

A similar shift occurred in the photography industry. For decades, companies like Kodak dominated with film-based products. But as digital imaging emerged, many struggled to adapt. Kodak, ironically a pioneer in early digital technologies, failed to make the leap in time. Meanwhile, players like Canon and Sony seized the opportunity, leading the next wave of image capture and distribution.

Transitions of this kind demand more than vision; they require structure. To manage the uncertainty of innovation, many organizations rely on frameworks such as the *Technology Readiness Levels (TRL)* model. Originally developed by NASA, the TRL framework offers a systematic way to evaluate how mature a technology is at any given stage of development. Rather than leaping blindly from lab to market, companies can pace their investments, adjust expectations, and manage risk as a technology advances through successive phases.

The journey begins at TRL 1, where basic principles are observed. At this stage, scientific research is still highly theoretical. As the concept solidifies, TRL 2 involves early feasibility analysis and definition of the potential application. Moving into TRL 3, the first experimental validation occurs, typically through laboratory-scale demonstrations that confirm the concept is viable. By TRL 4, these concepts begin to coalesce into functional prototypes that are tested under more controlled but realistic conditions.

Progressing further, TRL 5 and TRL 6 involve increasing levels of integration and environmental testing. Technologies are no longer just working in isolation; they must now perform as part of a system and meet real-world performance criteria. At TRL 7 and TRL 8, the technology is deployed in operational settings and tested at scale. Every aspect, from reliability to interoperability, is scrutinized. Only at TRL 9 is the technology deemed market-ready, having proven its value under actual operating conditions and across full deployment scenarios.

Consider SpaceX's reusable rocket program. Early tests of Falcon 9's booster landings corresponded to TRL 5 and 6—demonstrations in representative environments. Over time, with repeated successful landings and refinements, the program reached TRL 9. What once seemed improbable—reusing rockets—became standard practice, dramatically reducing launch costs and opening new possibilities in space logistics.

Of course, TRL transitions are neither automatic nor easy. Each step requires resources, expertise, and strategic patience. The effort is justified by the outcome: a pathway to take a fragile, emerging idea and shepherd it into a robust, scalable, and commercially viable innovation.

In today's fast-moving environment, organizations must be adept not only at riding the current S-curve but also at identifying when to leap to the next. The interplay between incremental advances and radical shifts—when supported by structured frameworks like TRL and informed by S-curve dynamics—enables companies to manage risk, seize opportunity, and shape the future rather than merely react to it.

Managing Disruptive Technologies: Contrasting Approaches of Large Enterprises and Innovation-Driven Enterprises (IDEs)

Entrepreneurial ecosystems flourish in healthy social and economic environments, and among the multitude of startups they generate, a special subset stands out: innovation-driven enterprises or IDEs. These are not ordinary small businesses. They are ventures born with an ambition to transform, disrupt, and reshape entire markets through the power of breakthrough innovation. Operating in what we've defined as the Big-I space, IDEs tackle entirely new problems by crafting radical, often paradigm-shifting solutions. Unlike firms built on incremental improvements or regional market strengths, IDEs are born global. They aim high and wide, fueled by an appetite for risk and a bold vision of the future.

To bring their vision to life, IDEs must engage deeply and intentionally with a broad network of stakeholders. This includes not only customers and suppliers but also academic institutions, research centers, competitors, and government agencies. These collaborations are not merely supportive; they're foundational. IDEs rely on shared learning, rapid feedback, and access to resources that can only be mobilized through an open, connected innovation environment.

This operating model stands in sharp contrast to that of traditional SMEs. As Figure 5.2 illustrates, the distinctions between these two types of enterprises go far beyond scale. Small and medium-sized enterprises often pursue incremental innovation to defend or enhance their competitive advantage in local or regional markets. Their growth tends to be steady, linear, and closely tied to reinvested profits. IDEs, by contrast, enter the scene with disruptive intent. They take aim at entrenched industries and mainstream solutions, introducing radically new technologies or business models. They also embrace a far riskier financial model: negative cash flows are often expected in the early stages, with the understanding that if the venture succeeds, the returns can be enormous.

Feature	SMEs	IDEs
Market Focus	Focus on addressing local/regional markets	Focus on global markets
Small-I vs. Big-I Focus	Small-I innovation aims at sustaining competitive advantages	Big-I innovation (problem-solution space model) aims at challenging the mainstream
Job Creation	Create "lifestyle" jobs, generally performed locally (e.g. transformation industry, service industry, etc.)	Create "desirable" jobs, not necessarily performed locally
Equity Structure	Are most often businesses with very little external capital	Have fragmented equity, including wide array of risk capital providers
Growth Dynamics	Typically grow at a linear rate, investments tend to pay dividends	Start with negative cashflows, but if successful investments pay HUGE dividends

Figure 5.2 SMEs vs. IDEs. (From Majava, J., Kinnunen, T., Foit, D., & Kess, P. (2017). Analysing innovation-driven enterprises' stakeholders in two spatial ICT ecosystems, *International Journal of Management and Enterprise Development (IJMED), 16(3).* With Permission.)

The structural differences go deeper. IDEs usually operate with fragmented and sophisticated equity structures, often backed by a mix of venture capital, angel investors, and institutional funds. This investment diversity is necessary to sustain the extended development periods and experimentation cycles that Big-I innovation typically requires. SMEs, on the other hand, are more likely to rely on internal financing or modest lines of credit. Their innovation models, while valid and valuable, rarely aim to disrupt entire markets.

The journey of IDEs is also closely intertwined with disruptive technologies—innovations that at first may seem inferior to existing solutions in mainstream markets but eventually overtake them due to their potential for cost reduction, improved performance, or radically new capabilities. This process is clearly illustrated in Figure 5.3. The diagram shows how a disruptive technology, initially serving an underserved or niche market (line D), gradually improves in performance (curve C) and eventually enters and transforms the mainstream market (line B), displacing an incumbent technology (line A) in the process.

This pattern played out, for instance, in the hard drive industry in the 1980s. At the time, 5.25-inch drives were dominant in desktop computers. Then came the 3.5-inch drives—smaller, lighter, and cheaper but initially underperforming in storage capacity and speed. Their compact form factor,

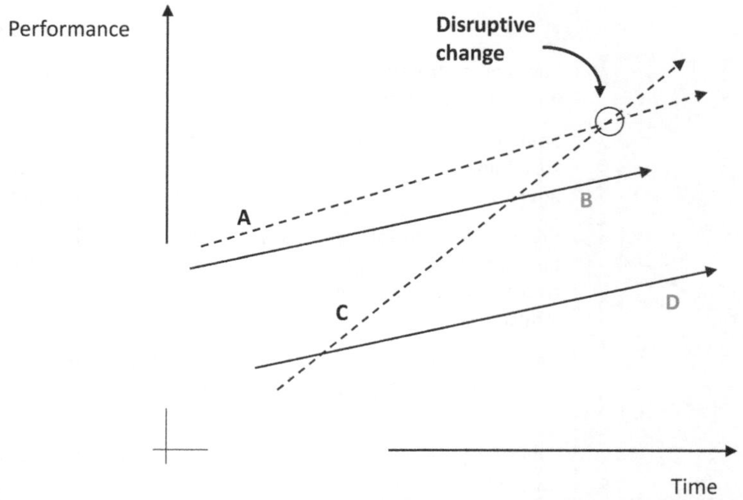

Figure 5.3 Disruptive technologies path.

however, made them ideal for laptops, a then-niche segment. As laptops grew in popularity, the performance of 3.5-inch drives improved, and eventually they displaced 5.25-inch drives even in desktops, becoming the new standard. It was a classic case of disruptive change, not because the new technology was better from the start, but because it improved faster and addressed emerging needs that incumbents overlooked.

The lesson for organizations is clear. Recognizing and navigating the path of disruptive technologies requires more than technical excellence. It demands strategic foresight and the willingness to invest in capabilities that may challenge current models and cannibalize existing business lines. This kind of transformation often meets resistance in large enterprises, where established routines, stakeholder expectations, and quarterly financial pressures tend to privilege short-term gains over long-term exploration.

And yet, for firms to stay relevant in the face of accelerating change, they must evolve. Organizational capabilities—defined by Dorothy Leonard-Barton (1992) as the intersection of resources, processes, and values that enable a company to deliver value—must be continuously reassessed and sometimes unlearned. When disruption looms, firms must be willing to break from what made them successful in the past in order to succeed in the future.

This is easier said than done. Research by Clayton Christensen and Michael Overdorf (2000) shows how difficult it is for established firms to simultaneously satisfy current stakeholders and prepare for radical technological shifts. The time horizons, cultures, and performance metrics involved are often in direct conflict. Due to stock market pressures, there is a tendency not to focus too much on converting the investment of time and resources into profits. However, breakthrough innovation often requires a longer investment period before yielding financial returns (Gilbert & Harris, 1984; Quinn, 1985). These dynamics make it challenging for large enterprises to pursue exploratory paths without compromising their immediate performance expectations. Yet, ignoring disruptive signals can be fatal. The very capabilities that once underpinned a company's competitive advantage may become liabilities when the environment shifts.

Ultimately, the contrasting approaches of SMEs, IDEs, and large enterprises to disruptive technologies are rooted in different attitudes toward risk, different time horizons, and different growth philosophies. IDEs push boundaries with speed and daring. SMEs build steadily and safely. Large enterprises must learn to do both—preserve what works while exploring what's next. Only then can they harness the power of disruption, not just survive it.

Large Enterprise Approach

As innovation reshapes entire sectors with increasing speed and unpredictability, large corporations find themselves under mounting pressure to keep pace. Many are now seeking ways to cultivate *dynamic capabilities* that not only fuel today's competitiveness but also can be sustained and renewed over time. At the heart of this challenge lies the tension between established organizational structures and the agility required to thrive amid disruption.

One of the keys to shortening the radical innovation lifecycle lies in how new ideas are developed and transitioned into market-ready solutions. When fresh thinking arises not only from internal teams but also from external collaborators—customers, suppliers, research institutions, startups—the path to commercialization becomes more fluid. In this context, *interactive innovation models* that link early-stage ideation to real-world market feedback loops are gaining traction. They help organizations move from abstract invention to validated opportunity faster, and more strategically.

For large enterprises, however, this shift is easier said than done. Disruptive changes often expose their most entrenched weaknesses: inflexible hierarchies, sluggish decision-making processes, and overly linear execution models. As Murray and Budden (2018) point out, these organizations are structurally disadvantaged when it comes to managing the ambiguity and speed of radical innovation. Their innovation processes are often optimized for efficiency, not exploration.

Christensen and Overdorf (2000) offered a now classic definition that remains highly relevant: *organizational processes are "the pattern of interaction, coordination, communication, and decision-making that employees use to transform resources into products and services."* This pattern, while highly effective in delivering established offerings at scale, can become a liability when organizations attempt to innovate beyond their core.

This is where the contrast with IDEs becomes particularly stark and instructive. IDEs thrive precisely because they are built for exploration. Their modus operandi revolve around a series of iterative innovation loops, not a single, monolithic development cycle. In these loops, each step involves hypothesis testing, controlled experimentation, and rapid evaluation. Learning happens continuously. Risk is incrementally absorbed. Resources are committed gradually, as understanding deepens and uncertainty recedes.

By contrast, large enterprises have historically operated within a single-loop model—designing and building the full solution upfront, only to discover late in the game whether it meets customer needs or not. This approach, while once reasonable in more stable markets, is now fraught with risk. Customer expectations evolve rapidly; competitors move quickly. By the time the solution is deployed, the problem it was designed to solve may have morphed, or disappeared entirely.

This divergence in approach reveals both a challenge and an opportunity for large corporations. On one hand, their organizational DNA can inhibit responsiveness. On the other hand, partnerships with IDEs offer a vital source of renewal. By engaging with innovation-driven firms—whether through venture arms, accelerators, or co-development models—large enterprises can tap into faster innovation cycles and alternative ways of thinking. More importantly, they can begin to internalize some of these practices themselves.

Ultimately, the goal for large enterprises is not to mimic startups but to blend the discipline of scale, in terms of entrepreneurial capabilities, with the agility of exploration, in terms of innovation capabilities. That requires rethinking how innovation is structured, resourced, and governed. It demands new incentives, new collaboration mechanisms, and, critically, a new mindset—one that embraces uncertainty not as a threat but as a source of strategic possibility.

A Corporate Pharmaceutical Example of Innovation Processes

The pharmaceutical industry offers one of the clearest and most structured illustrations of how large enterprises manage innovation on a grand scale.

Pharmaceutical Company innovation process

	Idea Generation			Idea Conversion			Implementation		
Innovation Framework	IN-HOUSE Creation within a unit	CROSS-POLLINATION Collaboration across units	EXTERNAL Collaboration with parties outside the firm	SELECTION Screening and initial funding	DEVELOPMENT Movement from idea to first results	DIGGING Movement from first results to solution	DEPLOYMENT Movement from results to industry	MARKETING Go-to-market process	LIFE-CYCLE Value creation extension
'Pharma' Innovation Processes	PRE-DISCOVERY Understanding the 'disease' and choosing a valid target molecule	DRUG DISCOVERY Project endorsement in designing a prototype of a drug molecule	NETWORKING Biotech firms and public sector institutions engagement	CAPITAL RATIONING	PRE-CLINICAL Investigational New Drug (IND) Submission	CLINICAL TRIALS 'Phase 1, 2, 3' and New Drug Approval (NDA) Submission	SCALE-UP FOR MANU-FACTURING	COMMER-CIALIZATION	POST-MARKET MONITORING AND 'Phase 4' TRIALS
ideas	from INCEPTION →								→ to IMPACT

Figure 5.4 Large enterprise innovation process framework: An example. (From Hansen, M. T., Birkinshaw, J., (2007). With Permission.)

As shown in Figure 5.4, the innovation process unfolds in a disciplined journey that begins with the inception of an idea and culminates in its impact on the market. This journey is typically divided into three major sequential phases: idea generation, idea conversion, and implementation. Within each phase lies a series of distinct activities—each involving different stakeholders, requiring tailored resources, and guided by specific performance questions.

The idea generation phase captures the early, often ambiguous stages where potential solutions begin to take form. It starts with "in-house creation," where internal R&D teams engage in pre-discovery work—mapping disease targets, identifying potential mechanisms, and exploring viable molecules. This stage lays the scientific groundwork, but it also raises a fundamental question: is the R&D function genuinely fostering open-ended inquiry, or is it subtly biased toward familiar paths?

To complement internal ideation, many pharmaceutical companies actively encourage "cross-pollination" between departments—R&D, marketing, legal, and finance—to align scientific curiosity with strategic viability. For instance, input from marketing can shape the commercial potential of a molecule, while legal and finance teams assess regulatory feasibility and cost implications.

But innovation rarely flourishes in isolation. Increasingly, companies are looking beyond their walls, collaborating with biotechnology firms, startups, and public research institutions to access novel ideas, scientific breakthroughs, and specialized capabilities. These external partnerships are more than opportunistic alliances; they are a strategic necessity. The key evaluative question becomes: Is the company effectively tapping into external sources of innovation to broaden its opportunity landscape?

Once promising ideas are identified, the process transitions to the idea conversion phase. Here, the goal is to turn exploratory concepts into viable candidates for development. This begins with "selection," where capital is allocated selectively to those ideas with the most promise. Senior leaders and cross-functional teams evaluate feasibility, commercial potential, and strategic fit. Rigorous screening and resource prioritization help weed out weaker projects early, but this phase also tests the organization's discipline: Is funding truly guided by merit or by internal politics and inertia?

Next comes "development," marked by preclinical studies and the submission of Investigational New Drug (IND) applications. Legal teams join forces with R&D to ensure that regulatory and safety protocols are satisfied. The process then deepens with "digging," the term used to describe the initiation of clinical trials. At this stage, the stakes rise dramatically: costs skyrocket, timelines extend, and the probability of failure looms large. A successful outcome here depends not just on scientific rigor but also on organizational alignment and execution speed. Submitting a New Drug Application (NDA) signals a pivotal milestone, often years in the making.

The final leg of the journey, implementation, focuses on transforming an approved drug into a product with real-world impact. "Deployment" involves scaling up manufacturing and ensuring quality control at industrial levels. Cross-departmental collaboration continues, now emphasizing operational excellence. Companies must assess their own agility: Are manufacturing and supply chain systems capable of meeting demand without delays or cost overruns?

This leads into "marketing," where legal, finance, and commercial teams craft the product's market entry strategy. A successful launch hinges not only on product efficacy but also on messaging, positioning, and timing. In this phase, the evaluative lens shifts to market performance: are commercialization efforts truly adding measurable value?

Finally, "lifecycle management" takes center stage, encompassing post-market surveillance, customer feedback, and *Phase 4* trials. With intellectual property protection windows typically lasting 10–15 years, companies are under immense pressure to extract maximum value before patent expiration. Monitoring real-world outcomes, managing safety risks, and sustaining relevance are all vital to prolonging a product's shelf life and ROI.

The stakes in pharmaceutical innovation are uniquely high, with development costs often exceeding $1 billion and enormous associated risks. Failure rates remain dauntingly high, and time is a constant constraint. Many innovations never reach the market—not because the science was flawed but because the complexity of the process proved insurmountable.

To navigate these challenges, the industry has gradually evolved toward what's known as the *"drug-centered trifecta model"* (Petrova, 2013). This approach centers on a strategic ecosystem forged by three complementary forces: large pharmaceutical corporations, biotechnology firms, and public research institutions. Each plays a specialized role—pharma giants bring scale and commercialization muscle, biotech companies drive applied R&D and molecule discovery, and public institutions generate the foundational science. Together, they mitigate risk, reduce inefficiencies, and accelerate the pathway from molecules to market.

In sum, the pharmaceutical innovation process, though resource-intensive and complex, offers a compelling model for how large enterprises can structure innovation at scale. It is not just about moving from idea to impact; it is about asking the right questions, fostering the right partnerships, and continually managing uncertainty. In such a demanding and regulated environment, only those companies that treat innovation as a multi-stakeholder, end-to-end system will emerge with a sustainable competitive edge.

Innovation-Driven Enterprises (IDEs): A Dynamic Innovation Model

IDEs don't follow a linear script. Instead, their growth unfolds through a continuous process of cycles—each one a deliberate experiment that tests a different dimension of the problem-solution fit. Rather than waiting for long-term certainty, IDEs move quickly. They generate evidence, validate assumptions, and reduce risk through iteration. As each cycle yields new insights, confidence grows, and so does the interest of investors—who step in gradually, injecting more capital at successive stages.

This approach hinges on two fundamental phases: experimentation and evaluation. In the experimentation phase, teams formulate a precise hypothesis—often focused on a specific customer pain point or opportunity. They then design lean, focused experiments that put this hypothesis to the test in real-world settings. The objective is not perfection, but speed. Rapid, actionable insights are far more valuable than elaborate but untested ideas.

Once the evidence is gathered, the evaluation phase takes over. Here, every data point is scrutinized. Does the evidence support the hypothesis? Does it warrant pivoting or doubling down? These evaluations guide the next move, setting the direction for either refinement or escalation.

From Linear to Iterative: Traditional Corporate Innovation vs. IDEs

This contrast is captured in Figure 5.5, which illustrates the difference between traditional corporate innovation and the approach favored by IDEs. Large enterprises tend to operate within structured, linear innovation models where ideas are developed through defined stages toward market entry. These models offer clarity and checkpoints, but they are often slow. In fast-evolving environments, they risk falling out of sync with market shifts or being disrupted by faster-moving rivals.

IDEs, by contrast, embrace a looping innovation framework: an iterative process of experimentation-validation cycles. This process is marked by milestones that are dynamic, not fixed. They pivot, refine, and re-test ideas in response to new information. This looping process doesn't just cut down the time to market; it builds agility into the organization's DNA. As a result, IDEs are better equipped to handle uncertainty, seize fleeting opportunities, and respond to competitive threats with speed and flexibility (see section "Experimental Innovation Loops: Reducing Uncertainty and Entrepreneurial Risk" in Chapter 4).

In Figure 5.5, the innovation loop milestones are intentionally drawn as a widening shape rather than a rectangular stage sequence. Early in the cycle, investment levels are deliberately kept small, while the probability of failure remains high. As uncertainty is progressively reduced through experimentation and validation, later milestones require materially larger investments, but the residual risk is comparatively lower. This risk investment profile aligns closely with the logic of venture capital, where capital is deployed in successive rounds, increasing exposure only when evidence supports scalability and value creation.

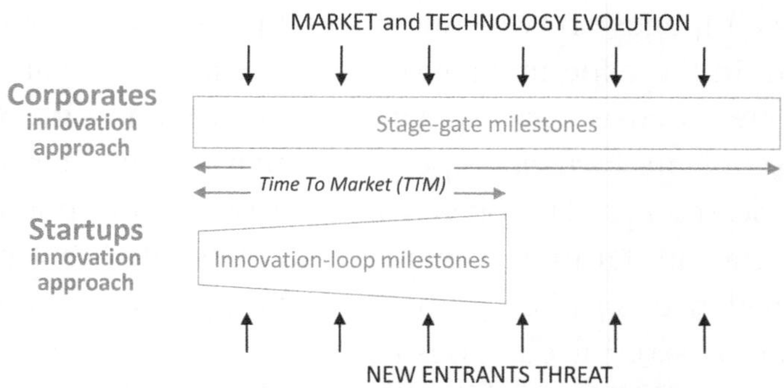

Figure 5.5 Large Enterprises vs. IDEs innovation process framework.

Adopting Lean and Agile Methodologies

The Lean Startup methodology provides the operating manual for this kind of iterative innovation. It revolutionized the way new businesses are launched and innovation is managed inside established firms. Traditional business planning—rooted in long-term forecasting and fixed execution paths—often proves ineffective in environments marked by rapid change. Lean Startup counters this by embedding agility, learning, and customer feedback at the core of its process (Blank, 2013).

Figure 5.6 helps visualize this transformation. It contrasts Lean Startup and traditional innovation models across seven key operational dimensions: from strategic intent and engineering practices to organizational structures and responses to failure. Lean Startup favors a hypothesis-driven mindset. Instead of dictating a rigid business plan, it sketches out value hypotheses and then tests and adapts them through direct customer interaction. It champions organizational agility, relies on different success metrics (such as customer acquisition cost), and treats failure not as a problem but as a necessary ingredient for learning.

	Lean	Traditional
Strategy	Business Model Hypothesis-driven	Business Plan Implementation-driven
New Product Process	Customer Development Get out of the office and test hypothesis	Product Management Prepare offering for market following a linear, step-by-step plan
Engineering	Agile Development Build the product iteratively and incrementally	Agile of Waterfall Development Build the product iteratively, or fully specify the product before building it
Organization	Customer and Agile Development Teams Hire for Learning, nimbleness and speed	Departments by Function Hire for experience and ability to execute
Financial Reporting	Metrics that Matter Customer acquisition cost, lifetime customer value, churn, viralness	Accounting Income statement, balance sheet, cash flow statement
Failure	Expected Fix by iterating on ideas and pivoting away from ones that don't work	Exception Fix by firing executives
Speed	Rapid Operates on good-enough data	Measured Operates on complete data

Figure 5.6 Lean Startup vs. traditional methodology.

This philosophy manifests through tools like the *Business Model Canvas*, which replaces exhaustive business plans with concise, testable hypotheses about customer needs, value propositions, channels, and revenue streams. These assumptions are continuously validated through *Customer Development*, an iterative process that brings potential users into the innovation loop from day one. If the hypothesis falters, IDEs pivot. They do not stubbornly stick to broken models; they evolve them.

Agility, is also reflected in engineering and product development practices. Traditional product development follows a phased, waterfall structure. Lean flips that. It uses agile development cycles to build a Minimum Viable Product (MVP), a lean prototype designed for rapid market testing and iterative refinement (see section "Balancing Speed, Credibility, and Competitive Space in MVP Design" in Chapter 2). The focus is on speed, simplicity, and adaptability. Tools like open-source software, cloud platforms, and contract manufacturing make this feasible at a fraction of the traditional cost.

Once the product-market fit is validated, IDEs scale. This shift from exploration to execution involves formalizing teams, structures, and systems. Yet, the core learning engine—the loop of experimentation and customer feedback—remains intact. This dual ability to explore and execute is at the heart of what ISO 56001 now seeks to institutionalize. The standard promotes repeatable innovation, structured processes, and a culture of continuous improvement. For organizations, this translates into fewer failures, faster time to market, and processes built for constant renewal (Blank, 2013).

However, Lean is not a cure-all. Its strength lies in structured environments where process efficiency and cost control dominate. In more volatile, ambiguous contexts, Lean may fall short (Welo, 2011). Here, *Agile methodologies*—with their emphasis on adaptability, collaboration, and rapid customer response—excel (Hannola et al., 2013).

Problems emerge when Lean is applied rigidly. Over-standardization can strangle creativity, reduce learning, and, ironically, increase costs (Johnstone et al., 2011; Browning & Sanders, 2012; Schulze & Störmer, 2012). The ideal, therefore, is a synthesis: a *"LeAgile"* approach that fuses the structure of Lean with the flexibility of Agile. This hybrid model allows organizations to remain efficient without sacrificing responsiveness (Wang et al., 2012).

Implementing LeAgile requires skilled leadership—people capable of reading the context and adjusting the mix accordingly. The challenge lies in striking the right balance between operational control and strategic fluidity (Magnusson, 2000; Putnik & Putnik, 2012).

Selecting and Structuring Innovation Approaches

Navigating the innovation journey requires more than just creativity; it demands conscious choices at every stage to ensure competitive relevance. This begins with selecting methodologies that not only match the phase of innovation but also align with the organization's strategic intent.

One powerful example is Gary Crocker's[1] approach to medical innovation, which illustrates a comprehensive, four-stage model of innovation. It starts with the *recognition of unmet market needs*—insights often gleaned not from spreadsheets but from observing the world with curiosity and empathy. This early exploration is followed by the *problem definition* phase, where innovators are urged to pause and reflect. Rather than rushing into solution mode, they are encouraged to deeply understand the nuances of customer pain points. As Mike Maples Jr.[2] demonstrates in his transformation of technical support services, prematurely assuming a solution can be a costly misstep. Instead, real value begins with framing the right problem.

The third stage, *rapid prototyping*, is about giving shape to ideas through theoretical constructs, virtual models, and, eventually, MVPs. It is a critical moment when innovation moves from concept to market interaction. Here, startups like Coin—a smart card and companion app designed to consolidate multiple credit, debit, and loyalty cards into a single physical device—offer compelling lessons: by putting simplified yet functional versions of their product into users' hands, they were able to validate assumptions with real-world data, not guesswork.

Beyond product design, the heart of sustained innovation often lies in business model experimentation. As Furr and Dyer (2014) argue, innovation strategies must be embedded within the business model itself. This means questioning not just *what* we offer but *how* we create, deliver, and capture value. Only by aligning strategic experimentation with real-world feedback can organizations build products that resonate.

Business model experimentation should, of course, take into account a critical distinction between incremental and disruptive innovation. Incremental innovation thrives in familiar territory, relying on bottom-up processes and conventional metrics to refine existing offerings. But disruptive innovation is a different beast; it demands bold, top-down leadership, evolving KPIs, and a safe space where radical ideas can be nurtured without being immediately shot down by corporate antibodies, a phenomenon sometimes dubbed the "bazooka syndrome." As Govindarajan and Sebell (2012) note, successful disruption often hinges not just on the quality of the idea but on an organization's willingness to protect and guide it through early fragility.

Evaluating the effectiveness of such innovation initiatives requires deliberate design. Organizations can assess innovation at multiple levels: within a single unit, across the enterprise, or within broader ecosystems that span partners, suppliers, and other stakeholders. In early phases, it is often advisable to start with smaller, well-understood organizational areas. This localized approach enables controlled experimentation, allowing internal teams to test methods, calibrate metrics, and learn quickly before scaling. Evaluation can combine qualitative techniques—such as interviews, surveys, and structured workshops—with targeted external input, for example expert reviews, advisory panels, or benchmark comparisons that add an independent perspective.

One of the most widely adopted frameworks is the *capability maturity model*. This structured approach enables organizations to assess where they stand—and where they aim to go—across key innovation dimensions. These models typically span five maturity levels, ranging from ad hoc practices (zero capability) to fully optimized innovation systems. Not every area must strive for the highest maturity; rather, the goal is to define strategic priorities and evolve capabilities where they matter most. For instance, achieving Level 3 often signals a proactive and well-governed approach, even if some elements remain aspirational.

More broadly, innovation performance assessments play a vital role in transforming raw ideas into market-ready solutions. They help ensure that innovation efforts are not isolated initiatives but part of a coherent, feedback-driven system aligned with overarching strategic goals. The insights gathered during assessments should inform not only tactical decisions but also guide broader portfolio evaluations and innovation governance.

To support such evaluations, organizations must embrace a systemic mindset—one that integrates strategic vision, leadership commitment, and collaborative infrastructures. Innovation thrives not through isolated brilliance but through ecosystems that are robust, adaptable, and capable of sustaining change over time. As Karlsson and Magnusson (2019) remind us, success lies not in assembling an assortment of tools but in orchestrating them harmoniously to serve clear, evolving objectives.

At their best, these assessment practices do more than measure progress—they convert experience into insight. In other words, evaluation is only valuable if it accelerates organizational learning and feeds better decisions into the next innovation cycle.

Transforming a stagnant organization into an innovation leader requires one thing above all: learning—not as an abstract ideal but as an operational discipline that can be designed, measured, and reinforced. A five-principle framework supports this journey: frequent iterations with real-time feedback; simplified models to accelerate learning; diverse learning styles (e.g., storytelling, prototypes, data); small, nimble teams; and side-by-side comparison of multiple options to refine the best path forward.

The *NABC Value Proposition Framework* embodies this logic. NABC stands for Need, Approach, Benefits, and Competition: every innovation must address a real need, articulate a defensible approach, deliver measurable advantages, and outperform rivals. Champions—those individuals who relentlessly pursue new solutions—are supported by structured forums designed to sharpen and test these ideas. Traditional innovation often fails, as evidenced by low patent-to-market ratios and loss-making university incubators. But these limitations can be overcome. With active learning, iterative validation, and frameworks like NABC, even legacy institutions can unlock scalable innovation and long-term value (Carlson, 2020).

Ultimately, innovation is not a one-size-fits-all process. It requires ongoing calibration between goals, capabilities, and methods. It calls for a blend of structure and flexibility, vision and iteration, and analysis and intuition. And most of all, it calls for leaders who are willing to invest in long-term value creation—leaders who recognize that sustaining innovation is not just about what we invent but how we nurture, test, and scale it across time.

Considering all this, sustainable innovation is not a single stroke of genius; it's a progression. It begins with sourcing and generating ideas, while identifying constraints early. Legal, technical, and integration factors must align with customer and business objectives. Afterward, ideas are tested through structured hypothesis validation before moving into incubation. Here, MVPs are refined, user data is collected, and technical debt is managed to enable scalability. This kind of disciplined innovation—when executed properly—translates into successful market entries and long-term scalability (Blank & Newell, 2017).

Measuring Innovation Performance: A Balanced, Strategic, and Adaptive Approach

In today's volatile and fast-evolving business landscape, measuring innovation performance is no longer a luxury; it's a strategic imperative. Organizations striving to sustain their competitive edge and demonstrate the value created by innovation must design assessment systems that are both structured and adaptable. An effective innovation measurement approach doesn't just provide clarity; it aligns innovation activities with strategic intent, helping leaders make informed decisions, while continuously learning and improving.

To be truly meaningful, innovation measurement should span three interdependent levels: the innovation processes, the innovation portfolios, and the overall IMS. At each level, different types of indicators are needed to capture what matters most.

At the innovation process level, performance is evaluated using a combination of qualitative and quantitative indicators across four core measurement dimensions.

First, *inputs* gauge the volume and quality of resources invested. These may include monetary investment, specialized human capital, and the depth of external collaborations.

Throughputs secondarily track the pace and vitality of innovation activities. This includes the frequency of ideation sessions, customer co-creation workshops, expert consultations, technology scouting, and the cadence of prototyping and experimentation.

Furthermore, *outputs* reflect the tangible results: the number and novelty of solutions generated, the speed of market delivery, and resource utilization efficiency.

Finally, *outcomes*, perhaps the most strategic layer, assess the broader impact of innovation—improvements in customer satisfaction, sustainability metrics, market share expansion, brand equity, and the financial returns attributable to innovation.

Organizations may adopt streamlined scorecards—perhaps selecting no more than three key indicators per dimension—to preserve clarity and strategic focus. The chosen indicators should reflect the ambition of the innovation being pursued, whether incremental, semi-radical, or radical in nature.

When moving to the innovation portfolio level, measurement becomes a matter of balance, coherence, and strategic foresight. Indicators should assess not just the volume of innovation activities but their alignment with risk profiles and market opportunities. Relevant metrics include the total allocation of resources (human, financial, technological) toward innovation; the innovation mix, often guided by the "70–20–10" heuristic (70% incremental, 20% semi-radical, 10% radical); and the expected returns across the development lifecycle, using not only standard financial metrics like NPV but also broader strategic value assessments.

Here, strategic dialogue becomes essential: decisions should weigh the risks of investing in bold ideas against the even greater risk of missing future opportunities.

Frameworks such as the ISO 56002 (ISO, 2019) standard offer valuable guidance for evaluating innovation maturity and coherence at the overall IMS level. According to this standard, a robust performance measurement system should capture the interplay between resources, leadership, strategy, and processes, while also identifying capability gaps, non-conformities, and opportunities for improvement. In this context, adopting the Balanced Scorecard methodology can be particularly effective, providing a multidimensional view that blends financial, social, environmental, and cultural indicators to track both performance and stakeholder value creation (OECD/Eurostat, 2018).

This holistic and adaptive approach ensures that performance measurement is not a bureaucratic exercise but a tool for continuous improvement. By fostering transparency, shared accountability, and strategic alignment, organizations can adapt dynamically to changing conditions and emerging innovation opportunities.

As illustrated in Figure 5.7, innovation performance can be evaluated across three tightly linked domains. *Inputs* represent the starting point—the investments and enablers of innovation. These include R&D spending as a percentage of sales, the number of human resources dedicated to innovation, the size and diversity of idea pipelines, and the origin of these ideas (internal vs. external).

Process Effectiveness reflects the efficiency and agility of innovation execution. Key indicators include adherence to development timelines and budgets, patenting activity (filed, granted, and protected), the speed to market for new offerings, and the rate of significant innovation among launched projects.

And finally, *Performance Outcomes* focus on impact. They track the contribution of new products to revenue, success rates in hitting financial targets, ROI on innovation efforts (ROIC), customer satisfaction, time to

Inputs	Process Effectiveness	Performance Outcomes
• R&D spending (% of sales) • Human resources devoted to innovation • Pipeline of ideas/concepts • Number of R&D projects in active development • Percent of ideas/concepts from outside the firm • Ratio of ideas from inside/outside	• Development activities: - Percent hitting gates on time - Percent meeting quality - Guidelines • Patenting activity: - Number filed - Number commercialized - Percent covered by patents • Budget vs. actual: - Time - Cost/investment • Average time to market: - Number of new products launched • % of projects that are major improvements	• % of sales from new products in past n years • Success ratio (% of meeting financial goals) • Revenue growth • Return on investment in innovation (ROIC) • % of profits from new customers (or occasions) • % of profits from new categories • Average time to break-even/cash • Customer satisfaction • Profit growth due to new products/services • Percent of profits from new products in a given period • NPV of portfolio • Potential of portfolio to meet growth targets

Figure 5.7 Inputs, process effectiveness, and performance outcomes.

profitability, and even broader metrics like NPV of the innovation portfolio and its capacity to meet growth targets.

Together, these dimensions provide an integrated view of how innovation investments are transformed into tangible, value-creating outcomes. They serve as a dashboard for strategic decision-making, guiding resource allocation, identifying bottlenecks, and unlocking the full potential of innovation efforts.

The Complexity—and Necessity—of Measuring Innovation

Despite the value of structured frameworks, measuring innovation remains one of the most complex challenges for organizations. This is especially true in today's Industry 5.0 economy, where innovation is increasingly intangible, digital, collaborative, and distributed across platforms, partners, and ecosystems. The tools we inherited from the industrial era—such as patent counts or R&D intensity—are necessary but no longer sufficient.

Indeed, many companies struggle to replicate the success of innovation leaders like Apple or Google, not because they lack ideas or resources, but because they lack robust, coherent measurement systems (Richtnér et al., 2017).

Some organizations fall into the trap of using too many disconnected metrics, which overwhelms teams and diffuses focus. Others avoid measurement altogether, believing that innovation is too creative or unpredictable to be measured. Both extremes are detrimental.

To move forward, organizations must embrace a system-oriented approach—one that considers the full lifecycle of innovation from input to outcome. This includes assessing not only resources and activities but also how these translate into customer value and strategic progress (Anthony, 2013; Tucker, 2021).

The concept of Return on Innovation Investment (ROII) provides a helpful lens here. As proposed by Scott (2013), ROII dissects innovation performance into three key elements: *Innovation Magnitude*—the strategic or financial impact of successful innovations; *Innovation Success Rate*—the proportion of initiatives that deliver real value; and *Investment Efficiency*—how effectively time, capital, and talent are deployed relative to value created.

This framework allows organizations to manage a portfolio of innovations with different risk-return profiles, from safe, incremental gains to high-risk, high-impact breakthroughs.

The importance of strategic coordination in innovation measurement was also highlighted at the 2019 Mack Institute Fall Conference (Shimada, 2020). No single metric can fully capture innovation success. While financial indicators like the Vitality Index[3] or Holt Innovation Premium[4] are useful, they must be complemented by qualitative measures—creativity, team collaboration, or stakeholder engagement. For example, the Research Quotient (RQ) developed by Professor Anne Marie Knott and Carl Vieregger (2019) measures how effectively companies convert R&D spending into market impact, offering deeper insight into innovation capability.

Good measurement also demands attention to organizational behavior and culture (see section "Establishing a Culture for Innovation" in Chapter 6). Innovation thrives not just through structure and incentives but through curiosity, openness to risk, psychological safety, and a shared sense of purpose. A company that fosters a culture of experimentation and learning will often outperform one that focuses narrowly on metrics (Gann, 2019).

Simple indicators like *idea churn rate*—the share of ideas intentionally dropped early—can be surprisingly powerful in revealing the health of the innovation pipeline. Moreover, avoiding "metric fatigue" is key. Reporting should be consistent and meaningful but not overwhelming.

As innovation itself evolves, so too must our methods of measurement. Traditional indicators must be supplemented with new tools that reflect the digital, real-time, and insight-driven nature of Industry 5.0. Organizations are increasingly leveraging AI, advanced analytics, and real-time data streams to track emerging innovation patterns. For instance, AI can detect early innovation signals in user behavior or digital marketplaces, offering a proactive edge.

Finally, measuring innovation performance is both a science and an art. It requires discipline without rigidity, structure without bureaucracy, and insight without overcomplication. By integrating input, process, and outcome metrics, adopting strategic frameworks like ROII, and cultivating a culture that embraces learning and adaptability, organizations can gain visibility into their innovation engine—and continually tune it for better results.

Ultimately, performance measurement should not be a static dashboard but a dynamic compass—one that guides innovation journeys toward sustained value creation and long-term success.

Notes

1. Gary Crocker is primarily known as an entrepreneur and innovator in the medical and biotechnology industries. Among his most notable contributions is his leadership in developing advanced medical devices that address specific healthcare challenges. One well-known example is his involvement in founding Research Medical Inc. (1983), which developed disposable medical devices used in heart surgery, significantly reducing complications and improving patient outcomes.
2. Mike Maples Jr. is a prominent American entrepreneur and venture capitalist, widely recognized as a founding partner of Floodgate, an influential Silicon Valley venture capital firm known for its investments in early-stage startups. He is renowned for supporting and identifying successful startups in their initial phases, having invested early in notable companies such as Twitter, Twitch, Lyft, Okta, and Chegg.
3. The *Vitality Index* measures the percentage of a company's revenue that comes from products or services launched within a recent time frame, typically the past three to five years. It reflects how effectively an organization is generating growth through innovation and helps assess the commercial success of its innovation pipeline.
4. The *Holt Innovation Premium*, developed by Credit Suisse's HOLT team, estimates the portion of a company's market value that exceeds the value justified by its current profitability and asset base. This excess—called the "innovation premium"—is interpreted as the market's expectation of future growth driven by innovation. It essentially reflects how much investors are willing to pay today for the promise of future innovation.

References

Anthony, S. (2013). How to Really Measure a Company's Innovation Prowess. *Harvard Business Review.*

Blank, S. (2013). Why the Lean Start-Up Changes Everything. *Harvard Business Review.*

Blank, S., & Newell, P. (2017). What Your Innovation Process Should Look Like. *Harvard Business Review.*

Browning, T., & Sanders, N. (2012). Can innovation be lean? *California Management Review, 54(4),* 5–19.

Carlson, C. R. (2020). Innovation for Impact: Value Creation as an Active Learning Process. *Harvard Business Review.*

Christensen, C. M., & Overdorf, M. (2000). Meeting the Challenge of Disruptive Change. *Harvard Business Review.*

David Gann, M. D. (2019). We Need to Measure Innovation Better. Here's How. *World Economic Forum.*

Furr, N., & Dyer, J. (2014). Choose the Right Innovation Method at the Right Time. *Harvard Business Review.*

Gilbert, R. J., & Harris, R. G. (1984). Competition with lumpy investment. *The RAND Journal of Economics, 15(2),* 197–212.

Govindarajan, V., & Sebell, M. (2012). Match Your Innovation Process to the Results You Want. *Harvard Business Review.*

Hannola, L., Friman, J., & Niemimuukko, J. (2013). Application of agile methods in the innovation process. *International Journal of Business Innovation and Research, 7(1),* 84–98.

ISO (2019). ISO 56002:2019 – Innovation management system – Guidance. *International Organization for Standardization.*

ISO (2020). ISO 56000:2020 – Innovation management – Fundamentals and vocabulary. *International Organization for Standardization.*

ISO (2024). ISO 56001:2024 – Innovation management system – Requirements. *International Organization for Standardization.*

Johnstone, C., Pairadeu, G., & Pettersson, J. (2011). Creativity, innovation and lean sigma: A controversial combination? *Drug Discovery Today, 16(1,2),* 50–57.

Karlsson, M., & Magnusson, M. (2019). The Systems Approach to Innovation Management. In J. Chen, A. Brem, E. Viardot, & P. K. Wong (Eds.), The Routledge Companion to Innovation Management (pp. 73–90). *Routledge.*

Knott, A. M., & Vieregger, C. (2019). Does R&D drive growth. *Available at SSRN 2382885.*

Leonard-Barton, D. (1992). Core capabilities and core rigidities: A paradox in managing new product development. *Strategic Management Journal, 13(S1),* 111–125.

Magnusson, M. (2000). Innovation and Efficiency - A knowledge-based approach to organizing industrial firms. (Ph.D. dissertation). *Chalmers University of Technology.*

Murray, F., & Budden, P. (2018). Innovation Ecosystems: A New Approach to Accelerating Corporate Innovation and Entrepreneurship. *Sloan School of Management.*

OECD/Eurostat (2018). Oslo Manual 2018: Guidelines for Collecting, Reporting and Using Data on Innovation. 4th Edition. *OECD Publishing.*

Petrova, E. (2013). Innovation in the Pharmaceutical Industry: The Process of Drug Discovery and Development. In M. Ding, J. Eliashberg, & S. Stremersch (Eds.), Innovation and Marketing in the Pharmaceutical Industry: Emerging Practices, Research, and Policies (pp. 18–81). *Springer.*

Putnik, D., & Putnik, Z. (2012). Lean vs agile in the context of complexity management in organisations. *The Learning Organization, 19(3),* 248–266.

Quinn, J. B. (1985). Innovation and corporate strategy: Managed chaos. *Technology in Society, 7(2–3),* 263–279.

Richtnér, A., Brattström, A., Frishammar, J., Björk, J., & Magnusson, M. (2017). Creating Better Innovation Measurement Practices. *MIT Sloan Management Review.*

Schulze, A., & Störmer, T. (2012). Lean product development-enabling management factors for waste elimination. *International Journal of Technology Management, 57(1, 2, 3),* 71–91.

Scott, D. T. (2013). The New Rules of Lead Generation: Proven Strategies to Maximize Marketing ROI. *AMACOM Division of American Management Association.*

Shimada, A. (2020). Choosing Innovation Metrics for Success in the Digital Era. *Mack Institute for Innovation Management.*

Tucker, R. B. (2021). Innovation Can Absolutely, Positively Be Measured. Four Tips From The Pros For Doing It Right. *Forbes.*

Wang, X., Conboy, K., & Cawley, O. (2012). "Leagile" software development: An experience report analysis of the application of lean approaches in agile software development. *The Journal of Systems and Software, 85,* 1287–1299.

Welo, T. (2011). On the application of lean principles in Product Development: A commentary on models and practices. *International Journal of Product Development, 13(4),* 316–343.

Chapter 6

Organizational Culture and Effective Collaboration

Culture and Collaboration According to ISO 56000 Series Requirements

At the heart of any thriving innovation system lies a culture that doesn't just permit creativity—it actively nurtures it, alongside a disciplined drive for execution. According to the International Organization for Standardization (ISO) 56000 series, and particularly in *ISO 56001* (ISO, 2024), the message is clear: organizations must go beyond lip service and fully commit to cultivating a shared set of values, beliefs, and behaviors that champion openness to change, encourage intelligent risk-taking, and foster collaboration across functions and boundaries (ISO 56001, Clause 4.4.2).

This cultural foundation is more than a soft concept—it's a strategic lever. Traditional management tends to emphasize control, predictability, and efficiency. Innovation, on the other hand, flourishes in environments that tolerate ambiguity, welcome divergent thinking, and invite experimentation. The real challenge—and opportunity—lies in reconciling these two worlds. When an organization manages to create a cultural environment that supports both the creative birth of ideas and their structured implementation, it builds a bridge between exploration and execution.

The benefits of such a culture extend far beyond morale. First, it enables individuals and teams to explore unconventional paths, question long-held assumptions, and tap into their imaginative potential. Second, it allows for the accumulation of rich, collective intelligence through mechanisms

DOI: 10.4324/9781003743583-6

of sharing, co-creating, and testing. Third, it supports an agile execution mindset—one that balances focus with flexibility, and action with reflection. Crucially, when employees feel safe stepping outside their comfort zones and are supported in learning from setbacks, innovation becomes a living, adaptive process rather than a series of isolated initiatives.

This cultural dynamic is inseparable from the concept of collaboration, another cornerstone of ISO 56001. Innovation is rarely a solitary act. *Clause 4.4.3* explicitly calls for both internal and external collaboration, recognizing that the sharing of knowledge, capabilities, and resources across different stakeholder groups is a necessary condition for innovation to thrive. These groups can range from internal teams and business units to customers, suppliers, academic institutions, startups, and industry consortia. Each brings a unique lens, and it is at their intersection that breakthrough thinking often emerges.

Collaboration can take many forms. Open innovation networks help identify opportunities at the periphery of current business models, while more closed or protected environments allow for trust-building and deep co-development within key teams. The method chosen should align with the organization's innovation strategy, its maturity, and the nature of the opportunities it seeks to pursue. But regardless of structure, collaboration only takes root when there's a cultural willingness to listen, learn, and share beyond traditional boundaries.

In this cultural-collaborative equation, leadership plays an irreplaceable role (as introduced in Chapter 1, section "Future-Focused Leadership"). Leaders are not just figureheads—they are the active architects of the conditions under which innovation can flourish. Their task is to cultivate a workplace defined by openness, curiosity, and an unwavering focus on the end user. That means modeling feedback-seeking behavior, celebrating learning (even when it arises from failure), and dismantling hierarchical barriers that stifle initiative. Approaches like innovation hubs, self-directed teams, and corporate spin-offs can support this shift, allowing for faster decision-making and deeper engagement. At the same time, leaders must act as the stewards of strategic partnerships and cross-functional alliances, ensuring that the organization's knowledge base remains broad, integrated, and constantly refreshed.

Ultimately, combining a culture that embraces innovation with mechanisms that foster meaningful collaboration—internally and externally—enables organizations to navigate the dual challenge of imagining the new and making it real. This alignment ensures that innovative efforts are not random bursts of creativity but instead coherent, strategically guided pursuits that generate long-term value and resilience. ISO 56001 captures this vision well in *Clauses 4.4.2 and 4.4.3*, reminding us that innovation is not just about good ideas—it's about the structures and relationships that allow those ideas to take root, evolve, and scale.

Establishing a Culture for Innovation

The Interplay Between Strategy and Culture

There is a subtle, yet powerful interplay between an organization's strategic direction and its cultural fabric—one that is increasingly pivotal in fostering sustainable innovation. It is not simply that culture supports strategy; rather, strategy can actively shape and evolve culture. The ISO 56000 series, especially through Clauses 4.4.2 of ISO 56001 and ISO 56002 (ISO, 2019), affirms this duality: an innovation-oriented strategy provides vision and structure, while culture animates it, giving breath and energy to its execution.

A company that aspires to innovate must ensure its culture encourages openness to change, embraces smart risk-taking, and nurtures collaboration across silos. These shared values and behaviors are not decorative slogans. They are essential in reconciling the often contradictory demands of creativity and flawless execution. Yet this kind of culture doesn't emerge spontaneously. It must be crafted, refined, and above all, understood.

Edgar Schein's seminal model of organizational culture (2010) offers a compelling roadmap. He likens culture to a living system with three interdependent layers: at the roots lie individual values and assumptions; at the trunk, the visible norms and guiding principles of the organization, and in the leaves and branches, we find the outward behaviors, rituals, and expressions of that culture. For innovation to thrive, these levels must be coherent—aligned not just in theory but in the everyday decisions, language, and emotional tone of the workplace.

The process begins with self-awareness. Organizations must critically assess their existing cultural state—what people truly believe, how they behave under pressure, what gets rewarded, and what gets ignored. From there, attention can shift to the values that unify people in pursuit of something bigger than themselves. These values—empathy, curiosity, courage, openness, a sense of mission—serve as the backbone of innovation-driven cultures (Watkins, 2013).

Yet leaders often find themselves lacking a shared vocabulary for discussing culture, let alone data to guide its transformation. Culture remains intangible, elusive, resistant to quick fixes. It must therefore be treated as a mission-critical asset, requiring deliberate interventions such as awareness training, internal storytelling, coaching frameworks, selective recruitment aligned with innovation profiles, and benchmarking of the actual culture against the desired one (Denison, 1990; Makridis, 2019). Without this rigor, even the most inspired strategic intentions remain suspended in abstraction (see section "Strategic and Structural Leadership for Innovation" in Chapter 1).

Bridging the gap between a legacy culture and one that supports innovation demands more than slogans or values posters in the lobby. It requires interventions at every level of the organization. That might include rearticulating brand values, recalibrating KPIs to reward experimentation, or dismantling systems that quietly penalize risk. Through modeling new behaviors, restructuring teams, adjusting incentive systems, and giving visibility to early adopters, leaders begin to orchestrate alignment between the roots, trunk, and leaves of the cultural tree (Jaruzelski et al., 2011).

In fast-moving markets, where volatility is the norm, this kind of alignment becomes a strategic asset in itself. It makes organizations more adaptive, more resilient, and more capable of executing against future-facing visions. When culture and strategy dance in step, innovation is no longer a side project—it becomes part of the organization's DNA.

Transforming Organizational Culture for Innovation

When it comes to innovation performance, corporate culture often outweighs even the size of an research and development (R&D) budget. Research based on global executive surveys and longitudinal data reveals a striking truth: companies that align their innovation strategy with a supporting culture experience, on average, a 30% greater increase in firm value and a 17% boost in earnings over five years (Jaruzelski et al., 2011). And yet, fewer than half of organizations achieve this strategic-cultural coherence. The result? Wasted investment, frustrated teams, and underwhelming returns.

At the heart of cultures that foster innovation, four traits consistently stand out. The first is an unwavering commitment to deeply understanding and meeting customer needs—not just reacting to requests but anticipating latent desires. The second is pride—pride in the act of creation, in building things that didn't exist before. The third is openness: to new ideas, to failure, to perspectives that challenge existing mental models. And finally, a dedication to cross-functional collaboration—a willingness to break out of departmental boxes and build together across differences (Denison, 1990).

But recognizing these traits is just the beginning. The challenge lies in embedding them into the muscle memory of the organization. That means weaving openness and experimentation into how decisions are made, how leaders lead, how new hires are selected, and how performance is measured. It means creating systems of accountability that do not punish failure but reward principled risk-taking and learning. Culture, after all, is not static. It is fluid—shaped by incentives, stories, leadership behavior, and external shocks. Left unmanaged, it evolves in directions that may run counter to innovation (Watkins, 2013).

Makridis (2019) points to a revealing statistic: a one standard deviation (SD) improvement in organizational cultural variables such as clarity of communication and managerial quality is correlated with a 0.59 SD rise in revenue and a 0.71 SD gain in operating income. Perhaps even more telling, employees say they would forgo 1.7% of their salary for a better culture. Culture, in other words, is a form of capital—intangible but extraordinarily valuable.

So where does change begin?

It starts by observing what Watkins (2013) calls "critical incidents"—real-world situations that expose the tension between operations and innovation. For instance, when teams resist new ideas because they fear the short-term consequences on their performance metrics, they reveal a deeper cultural misalignment. Leaders must pay close attention to these moments, not to assign blame, but to understand the assumptions and power structures that give them life.

From here, the process becomes one of *dilemma articulation*. Leaders must surface the competing logics at play, drawing out both the rational and emotional stakes involved. By making these trade-offs explicit—and giving people a shared language to discuss them—they pave the way for meaningful debate, and eventually, deliberate cultural evolution.

Of course, not all dilemmas are created equal. Some—such as external regulatory pressures—are non-negotiable. Others, like outdated budget conventions, are ripe for challenge. The key is to prioritize based on strategic impact. Leaders must know where to focus their attention, where to intervene boldly, and where to allow incremental adaptation.

Once the critical dilemmas have been identified and addressed, it's time to find strategic compromises. This might mean trading short-term productivity for long-term product innovation or extending timelines to improve the quality of experimental work. These decisions often require complex coordination—across budgets, teams, and governance structures—and must be endorsed at the highest level of leadership.

But compromises alone are not enough. They must be institutionalized. Culture doesn't shift through conversation alone; it changes when new ways of working become the norm. That might involve restructuring performance reviews to reward curiosity, introducing venture-style funding for internal projects, or changing reporting lines so that R&D is supported by champions, not just supervisors. These visible changes send a powerful message: innovation is not optional—it's embedded.

The cultural transformation journey is not linear. It is cyclical and iterative. It requires constant observation of tensions, articulation of competing narratives, prioritization of trade-offs, and visible reinforcement of new behaviors. Over time, what once felt like an insurmountable contradiction—between efficiency and exploration—can become a dynamic tension that fuels growth. The most innovative organizations are not those that eliminate paradoxes but those that learn to harness them.

Enhancing Collaboration and Knowledge Sharing

Innovation Ecosystems and the Collaboration for Knowledge Integration

Why does innovation flourish in some places and not in others? This seemingly simple question opens up a complex, fascinating inquiry. Innovation, both historically and today, has never been evenly distributed. It gravitates toward specific hubs—dense, resource-rich, socially vibrant areas that become magnets for ideas, talent, and experimentation. Looking back, the Industrial Revolution of the 18th century began not by chance in Britain but in areas that boasted abundant coal, navigable waterways, skilled labor, and a burgeoning textile trade. Cities like Manchester in England—and later Manchester, New Hampshire—had the perfect confluence of conditions to host industrial booms.

Fast forward to the age of artificial intelligence (AI). One might expect that with today's ubiquitous connectivity, high computational capacity, and instant global communication, innovation would be more evenly distributed across the globe. Yet it isn't. Even in an era defined by decentralization and remote collaboration, innovation continues to cluster.

Why? Because innovation, at its core, is deeply relational. It thrives on proximity—not just physical but cognitive, institutional, and cultural. Indicators like patent filings, R&D expenditures, startup activity, and the formation of new ventures all reveal that, just like in the age of steam and iron, innovation today remains localized—even if the means of production and communication have evolved.

AI offers unprecedented technological possibilities, but it also introduces new organizational demands. It should be understood not merely as an efficiency tool but as a foundational force reshaping how industries operate, how decisions are made, and how value is created and delivered (see section "Strategic Context for Innovation and Technological Progress Over the Last Four Decades" in Chapter 2). To compete and thrive, companies must embrace open, fluid structures that encourage co-creation. Innovation increasingly depends on collaborative networks—an interconnected web of actors that includes universities, research institutions, government agencies, accelerators, consultants, and startups. Such ecosystems don't merely support innovation—they shape its very trajectory.

Crucially, this new era demands a new kind of leadership (as introduced in Chapter 1, section "Future-Focused Leadership"). Leaders must cultivate skills that go beyond operational excellence. They must become fluent in data, agile in cross-functional settings, and capable of guiding their organizations through uncharted, fast-evolving landscapes.

From a theoretical perspective, these collaborative dynamics were first captured by Everett Rogers in 1962, who framed innovation as a social diffusion process. Decades later, the concept of the *innovation ecosystem*—or iEcosystem—offers a more structured lens through which to understand these complex interdependencies. Rather than viewing innovation as an internal function or a linear process, the iEcosystem approach positions it as an emergent phenomenon shaped by multiple, interlocking drivers.

Increasingly, firms have recognized the limitations of closed innovation models. They've pivoted toward open innovation, inviting new ideas from external sources. But open innovation brings its own challenge: abundance. With too many ideas, companies face the paradox of excess—how to sort, prioritize, and act on what matters. This is where iEcosystems become essential: they offer a way to channel, not just generate, innovation.

Successful firms don't just open the floodgates. They focus their innovation efforts in specific geographies—often only one or two—where ecosystems are rich in talent, infrastructure, and knowledge flow. In fact, local iEcosystems have long existed. From the Renaissance guilds to 18th-century coffee houses and to Silicon Valley, these are dense, dynamic environments where relationships, ideas, and resources intertwine to create value.

At its heart, an iEcosystem is a network of interdependent actors—individuals, institutions, and organizations—that collectively enable innovation. It integrates two largely distinct but overlapping spheres: the *research economy*, which is driven by exploration, and the *commercial economy*, driven by application and scale.

Over time, the balance between these spheres has shifted. In the United States, since the 1970s, research and commercialization have grown apart. The result is a fragmentation of the iEcosystem. As corporate labs retreated from basic science, universities became the primary engines of discovery. Meanwhile, firms shifted their focus toward development and implementation. The connective tissue—historically embodied by institutions like Bell Labs or Xerox PARC—weakened.

Venture capital (VC) stepped in to bridge the gap between science and market, but only selectively, and mostly within high-yield industries like biotech, software, and energy. Research confirms this divide. A longitudinal study of over 4,600 U.S. firms engaged in R&D showed a 20% decline in scientific publications between 1980 and 2006. Meanwhile, recognition for corporate innovation, as measured by prestigious awards, also plummeted. In 1971, Fortune 500 companies accounted for 41% of such accolades. By 2006, they earned just 6% (Keller & Block, 2013; Arora et al., 2020).

Several emblematic cases illustrate this shift: Bell Labs was restructured under Lucent in 1996; Xerox spun off PARC in 2002; IBM under Gerstner in the 1990s reoriented its research toward market value; and DuPont shuttered its iconic Central R&D Lab in 2016. Today, large firms rarely aim to *discover*—instead, they aim to *translate*. Research occurs upstream, and invention happens in a more applied, collaborative setting.

This new division of labor has implications. University research is more likely to be groundbreaking but also more abstract. Corporate research is more utilitarian but often less visionary. For innovation to thrive, new mechanisms of collaboration must bridge the two worlds.

Firms engaging in iEcosystems must therefore select modes of collaboration carefully. Figure 6.1 maps these modes across two key dimensions: the

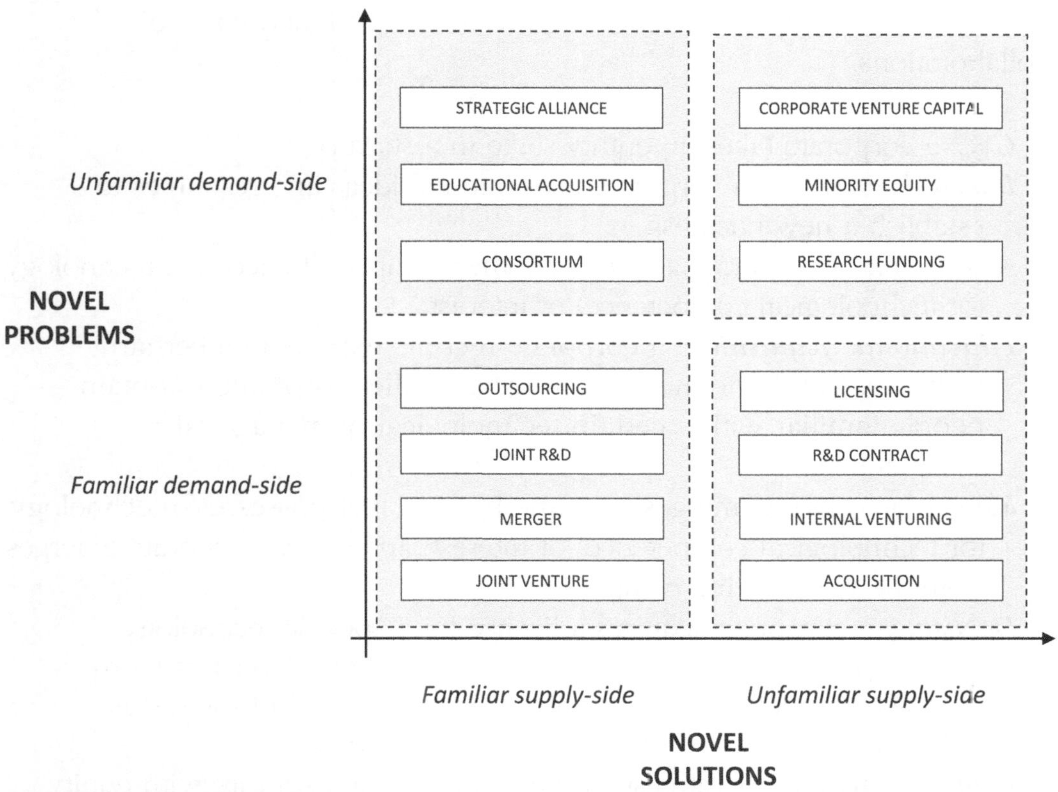

Figure 6.1 The map of collaboration types within iEcosystems at the innovation frontier.

familiarity with technological supply and the familiarity with market demand (Gardner & Ibarra, 2017).

In the lower-left quadrant of Figure 6.1—where both the problem and the solution are familiar—firms typically rely on structured, contractual collaborations such as outsourcing, joint R&D, mergers, and joint ventures. These arrangements are governed by clarity and trust, often formalized in detailed contracts with defined roles and exit strategies.

At the opposite extreme—the upper-right quadrant, where both supply and demand are unfamiliar—firms lean toward looser, more exploratory models like corporate venture capital (CVC), minority equity, or research funding. These rely more on mutual learning and trust than on predefined outcomes (Jackson, 2011).

In between lie hybrid models—such as strategic alliances, licensing, internal ventures, and acquisitions—which balance control and flexibility. These allow firms to innovate within their comfort zone while reaching into the unknown.

Here is a brief description of the previously mentioned types of collaborations.

CVC: a corporate takes an equity stake in a startup.
Internal venturing: a corporate uses internal ideas and resources to establish a new business.
Acquisition: a corporate acquires another company to access a technology (or technological competence) of interest.
Educational acquisition: a corporate recruits experts in a certain technological discipline or acquires a smaller corporate to obtain people familiar with a certain technological or managerial competence.
Merger: a corporate merges with another one that possesses a technology (or technological competence) of interest, and a new corporate emerges from the two existing companies.
Licensing: a corporate acquires a license for a specific technology.
Minority equity: a corporate buys equity in the source organization in which technology (or technological competence) of interest is embedded but does not have management control.
Joint venture: a corporate establishes a formal joint venture with equity involvement, and a third corporation is created, with a definite objective of technological innovation.
Joint R&D: a corporate agrees with others to jointly carry out research and development on a definite technology (or technological discipline), with no equity involvement.
R&D contract: a corporate agrees to fund the cost of R&D at a research institute, university, or startup for a definite technology.
Research funding: a corporate provides funding for exploratory research at a research institute, university, or startup to pursue opportunities and ideas for Big-I innovation.
Strategic alliance: a corporate shares technological resources with other companies to achieve a common Big-I objective (with no equity involvement).
Consortium: several companies and public institutions join their efforts to achieve a common objective of technological innovation (without equity involvement).
Outsourcing: a corporate externalizes technological activities and then acquires the relative output.

Each collaboration mode reflects a strategic calculation. For instance: CVC allows large firms to take equity stakes in startups; internal venturing draws from in-house talent to spin out new businesses; acquisitions provide immediate access to new technologies or capabilities; educational acquisitions recruit specialized talent through institutional links or smaller company acquisitions; joint ventures and mergers combine strengths across entities, whether formally or through equity involvement; licensing and outsourcing provide more transactional access to external capabilities; research funding, joint R&D, consortia, and strategic alliances emphasize shared discovery.

Figure 6.2 maps the stakeholders that constitute an iEcosystem. It distinguishes ten categories: universities, startups and innovation-driven enterprises (IDEs), government bodies, large enterprises, specialized VC firms, financial and legal institutions, innovation centers, cluster organizations, support ecosystem partners, and foundations. Each contributes to one or more of the iEcosystem's five driving forces, leveraging the *comparative advantages* summarized in the same framework (Bettanti et al., 2022).

These five forces are as follows: *Human Capital*—the engine of innovation, encompassing scientific talent, entrepreneurial drive, and cross-disciplinary skills; *Funding*—spanning public grants, philanthropic capital, angel investments, and corporate budgets; *Infrastructure*—both physical and

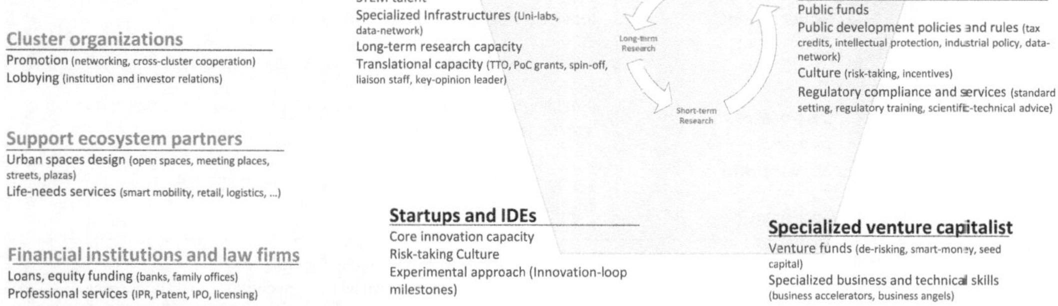

Figure 6.2 The iEcosystem stakeholder model. (From Bettanti, A., Lanati, A., & Missoni, A. Biopharmaceutical ecosystems: A stakeholder model and the case of Lombardy. The Journal of Technology Transfer. 2022. With Permission.)

digital, from labs to data platforms; *Specialized Demand*—informed customers and institutions that challenge innovators to solve real problems; and *Culture and Incentives*—the behavioral norms and motivational structures that shape collaboration and risk-taking.

The power of an iEcosystem lies in how these elements converge. At the core of the iEcosystem, entrepreneurs (startups and IDEs) introduce experimentation and often take leadership positions in initiatives. Universities contribute basic research, specialized laboratories, and graduate talent recruitment. Governments provide public grants and convening power for research initiatives. VC providers moving from early-stage angel investor networks to late-stage VC funds. They allocate funding at stages calibrated to technical and commercial milestones. Established companies offer scalability, established channels to market, and production capacity.

Finally, Figure 6.3 brings these dynamics into focus. It highlights the interplay between stakeholders and the ecosystem's core driving forces. For example, governments invest in infrastructure and incentivize

iEcosystem Driving Force	*Stakeholder*	*Role*
Human Capital (talent)	**University**	- Training the technical talent that contributes to innovation capacity (I-Cap) in an iEcosystem - Providing entrepreneurship education
	Corporates	- Shaping human talent
Funding	**Risk capital providers**	- Providing specialized funding
	University	- Spending on R&D
	Government	
	Corporates	
Infrastructure	**Government**	- Providing basic foundational infrastructure
	University	- Providing very specialized technical infrastructure
	Corporates	- Providing various kinds of space both for their own activities, but also for emerging entrepreneurs
Specialized Demand	**Government**	- Supporting demand around innovation capacity (I-Cap) - Engaging IDEs as suppliers
	Corporates	
Culture and Incentives	**Government***	- Providing the real sort of incentive foundation (*) cultural support comes from all the stakeholders in different ways

Figure 6.3 The iEcosystem driving forces. (From Budden, P., & Murray, F., Innovation Ecosystems: A New Approach to Accelerating Corporate Innovation and Entrepreneurship, Sloan School of Management, On-campus Program Presentation, Accessed March 15, 2018. With Permission.)

experimentation. Corporates bring demand, resources, and scale. Universities focus on research and talent development. Risk capital providers fuel the journey from prototype to product.

When these contributions are aligned, ecosystems can flourish. They create feedback loops: talent attracts funding, funding accelerates infrastructure, infrastructure enables experimentation, and experimentation inspires a culture of bold, persistent inquiry.

In such environments, innovation is not only possible—it becomes inevitable.

Engagement Tools for External Stakeholder Collaboration

Today, the orchestration of innovation requires more than traditional university-led research initiatives or direct startup acquisitions. The modern landscape demands a rich and varied toolkit of engagement mechanisms—tools that are not only participatory but also dynamic, repeatable, and inclusive of diverse stakeholders.

Organizations now approach innovation ecosystems with a broader repertoire. Beyond conventional R&D funding or mergers and acquisitions, firms experiment with more nimble and interactive formats: hackathons and hack days, open calls to collaborate, startup competitions, immersive visits to innovation hubs, co-creation with customers, thought leadership platforms, participation in entrepreneurial networks, and tailored R&D grants to academic institutions. These engagement mechanisms act as probes at the boundaries of innovation, helping firms explore emergent ideas, surface promising talent, and cultivate enduring relationships with early-stage ventures and creative individuals.

While this new engagement landscape is diverse, three formats have emerged as particularly effective and widely adopted: *hackathons*, *accelerators*, and *prize competitions*. These tools have proven effective not only as ideation platforms but also as vehicles for scouting talent, developing partnerships, and expanding innovation capacity across ecosystem boundaries (Gardner & Ibarra, 2017).

Figure 6.4 provides a structured comparison of these three mechanisms. It outlines their respective innovation goals, intended outcomes, modes of talent engagement, cost implications, and potential execution risks. But these are more than just formats—each tool embeds a different logic of collaboration, a different rhythm of interaction, and a different balance between openness and control.

	Hackathon	Accelerator	Prize Competition
Innovation Goal	Solve specific technical, business, or social issues	Better understand and leverage emerging technologies	New ideas to solve a well-known problem
Aim of the Tool	Collaboration, experimentation, and learning	Accelerate new venture creation	Attract and cultivate the best talent
Talent Engagement	Small teams	IDE cohorts	Individual IDEs
	Identify and recruiting talent externally	Education and mentoring	Direct engagement
	Ideas cross-fertilization	Bridging the corporate vs. IDE gap	Ideas exploration
Corporate Cost Comparison	- - -	Higher	- - -
Tool Risk in Execution	Weak coordination of talent teams	Too stringent control on IDEs cohorts	Poor engagement with IDEs individually

Figure 6.4 A comparison of the proposed engagement tools.

Take hackathons, for instance. These are fast-paced, time-boxed events where multidisciplinary teams converge around a specific technical, business, or social challenge. The emphasis is on rapid prototyping, creative friction, and experiential learning. The real value of a hackathon lies not only in the solutions developed but also in the ecosystem signals it surfaces—who shows up, how they work together, what ideas take shape, and which individuals demonstrate exceptional initiative or skill.

Consider the example of General Motors' global "AppShop" hackathons, run in collaboration with Chaotic Moon, TechCrunch Disrupt, and AppsWorld; these events offered more than just cash prizes. They provided developers with a space to co-create in partnership with the company, generating dozens of app prototypes tailored to the in-car experience. GM gained early insights, accelerated improvements to customer engagement, and unlocked novel revenue channels—all while strengthening ties with the developer community.

Accelerators, by contrast, offer a deeper, more structured engagement. These are intensive programs designed to help startups—often grouped into cohorts—scale rapidly through mentorship, training, funding, and exposure. Unlike hackathons, which are transient by design, accelerators are more immersive and longer term. They typically culminate in a "demo day," where ventures pitch to potential investors, corporate partners, and media.

The essence of an accelerator is knowledge transfer. Corporate sponsors benefit from early access to novel ideas, future investment opportunities, and the chance to shape technologies before they reach the market. Startups, meanwhile, gain entry into a curated network of mentors, technical experts, and customers. This model builds relational capital over time (Cohen, 2013). Importantly, accelerators are distinct from *incubators*, which focus more on protecting early-stage ventures by offering physical space and general support for an indefinite period. In contrast, accelerators are milestone-driven and performance-focused.

A complementary dynamic has emerged through *coworking spaces*—open, flexible environments where chance encounters often give rise to structured collaborations. These platforms, born from the ethos of openness and fluidity, provide fertile ground for cross-pollination between corporate teams and startup entrepreneurs (Spreitzer et al., 2015).

Meanwhile, prize competitions represent a third archetype—one that taps into the long-standing tradition of innovation challenges and incentive prizes. They engage a wide audience of problem-solvers, typically around a well-defined challenge. Prize competitions are effective for surfacing out-of-the-box thinking and tapping into niche or distributed expertise. Organizations that design these competitions carefully—establishing clear rules, relevant incentives, and robust evaluation criteria—can gather a portfolio of potential solutions while identifying individuals or teams worth nurturing further.

Success in such competitions is often gauged not only by the number or quality of submissions but also by what happens afterward: follow-on partnerships, the emergence of new business models, licensing agreements, or even spinouts. In this sense, prize competitions serve not just as contests but as open innovation pipelines.

Of course, each tool comes with its own trade-offs and execution risks, as highlighted in Figure 6.4. Hackathons can suffer from limited follow-through or lack of strategic alignment; accelerators may impose overly rigid frameworks that suppress spontaneous creativity; and prize competitions can generate fragmented engagement, especially when participants remain isolated from broader ecosystem structures.

Yet when these engagement tools are not used in isolation—but rather embedded into a broader ecosystem strategy—their combined potential becomes evident. The most effective organizations don't treat hackathons, accelerators, or competitions as one-off events. Instead, they integrate them with internal R&D efforts, align them with strategic objectives, and support them with the right governance models and cultural incentives.

In doing so, they transform episodic engagement into sustained innovation flow. They build long-term capabilities, not just short-term buzz. And perhaps most importantly, they create porous boundaries between the firm and the ecosystem—boundaries that let ideas, talent, and opportunity circulate freely.

When executed with care and vision, these engagement tools are not just instruments—they are catalysts. They breathe life into iEcosystems, bridging inside and outside, risk and reward, exploration and execution (Gardner & Ibarra, 2017).

Collaboration and Network Governance

Managing Corporate Partnerships

As shown in Figure 6.2, within today's iEcosystems, large firms operate in a delicate and paradoxical position—both as resource providers and as resource seekers. On the one hand, they bring capital, infrastructure, and distribution power to the table. On the other, they increasingly rely on external creativity and speed—elements they often lack internally—to complete the so-called innovation loop. What they offer in operational scale and process discipline, they frequently lack in agility, experimentation, and the tolerance for ambiguity that characterizes entrepreneurial ventures.

Established companies are, by design, optimization machines. They specialize in improving core functions, refining specialized processes, and scaling operations through complex, global supply chains. These strengths, however, can turn into liabilities when radical innovation is required. Their very structure—built for efficiency and predictability—can resist the uncertainty, iteration, and fast failure necessary for breakthrough advances (Chesbrough, 2003).

By contrast, startups or IDEs typically revolve around a single, untested idea. They are born to explore, pivot, adjust rapidly to user feedback, and learn by doing. They are unencumbered by legacy systems, annual planning cycles, or institutional inertia. This asymmetry is precisely what makes partnerships between large firms and IDEs both powerful and problematic.

In this context, large firms have become central actors in iEcosystems not merely by virtue of their scale but through their ability to curate access to real-world data and customer insight. By offering anonymized transaction logs, behavioral analytics, or market segmentation intelligence, they enable external innovators—whether startups or academic researchers—to build prototypes that are grounded in reality. These data-driven collaborations help direct creativity toward high-value problems and raise the odds of technological feasibility and user relevance (Jackson, 2011).

To engage effectively with IDEs, companies must move away from the outdated model of end-to-end innovation controlled by in-house labs. Instead, they must adopt a *portfolio mindset*—managing multiple, early-stage experiments before selecting a handful of strategic partnerships. This approach spreads risk and accelerates the pace of learning. It also demands cultural change: the willingness to accept failure not as a flaw but as part of the cost of discovery (Ries, 2011).

Still, tensions remain. Startups move at the speed of ambition. Corporates, by contrast, often move at the speed of budgeting cycles. The mismatch is acute: startup founders expect swift decisions and direct access; large companies rely on layered approval hierarchies, risk reviews, and fiscal gatekeeping. Valuable opportunities can be lost during lengthy due diligence processes. On the flip side, startups may frustrate their corporate counterparts by providing immature business cases, unclear customer segmentation, or a lack of data to support projected impact—all of which make integration into larger business units more difficult (Forrest & Martin, 1992; Slowinski & Sagal, 2010).

Bridging this gap requires firms to deploy their *core assets*—capital, talent, infrastructure, and distribution—in ways that are matched to the maturity of the partner venture. For example, experienced engineers might be embedded in startup teams to support design-for-manufacturing efforts. R&D budgets may be partitioned between non-dilutive grants and equity investments. Market access might be granted via "preferred supplier" agreements or joint go-to-market strategies. These interventions accelerate the development of external ventures while also reinforcing the firm's credibility as an ecosystem orchestrator (Hagedoorn & Zobel, 2015).

Importantly, startups are just one element of the broader partnership portfolio. Companies also engage with universities, public institutions, and VC networks—each of which requires its own governance architecture. Gone are the days of linear innovation models in which universities conducted basic research, passed it on for applied development, and then saw it commercialized by firms. Today, the relationship is circular, iterative, and deeply interdependent (Frølund et al., 2017).

Universities are no longer just producers of theoretical insight. They are talent pipelines, Intellectual Property Rights (IPR) generators, co-creators, and global knowledge hubs. Their value lies not only in laboratory breakthroughs but also in their ability to convene diverse disciplines, connect international research networks, and provide legitimacy to emerging technologies. In turn, businesses bring market awareness, funding, and scalability. Together, they shape each other's agendas—businesses influence academic priorities, while universities challenge corporate assumptions and broaden their innovation horizon.

Modern academic institutions are expected to go beyond their classical threefold mission of research, education, and public service. They must now embrace a fourth mission: *translational innovation*. This means cultivating the conditions under which basic science can be converted into usable, testable, and scalable applications. To do so, universities must build prototyping capacity, entrepreneurial education programs, and interdisciplinary accelerators that support real-world experimentation.

This requires a shift in pedagogy as well. In addition to disciplinary knowledge, students must be taught how to navigate uncertainty, pitch ideas, collaborate across fields, and understand market dynamics. Those trained in technical domains must also be fluent in entrepreneurship. Yet not all institutions are equally equipped for this transition. Engineering- and technology-oriented universities often adapt more naturally, while those with deep roots in classical liberal arts or theology may face steeper cultural and structural challenges.

For corporate leaders, engaging with universities is an exercise in timing and alignment. Long-term basic research partnerships can be used to explore frontier questions, while short-term, problem-focused collaborations may feed directly into product pipelines. Selecting the right entry point on the research-to-market continuum allows both sides to play to their strengths—universities providing exploratory depth and firms delivering operational speed and scale.

Government is another crucial partner in the iEcosystem, shaping the rules of the game and the incentives to play it. Governments define intellectual property regimes, competition policies, tax incentives, and procurement mechanisms. Increasingly, they are also expected to govern with agility, responding to fast-moving technological shifts with evidence-based regulations and adaptive funding programs. In this context, the private sector doesn't just receive policy; it helps shape it.

Companies, especially those leading in R&D and commercialization, are valuable policy advisors. They help governments calibrate tax credits, avoid innovation bottlenecks, and anticipate regulatory mismatches. However, governments must walk a careful line: supporting established firms without creating barriers for newer, entrepreneurial ventures. The goal is balance between enabling large-scale innovation and safeguarding the permeability that allows startups to emerge.

For firms, engaging with government actors requires strategic discernment. National ministries might focus on macroeconomic innovation agendas—such as smart specialization or defense tech funding—while regional and municipal agencies often concentrate on localized strengths. They might fund a university research center, sponsor an urban coworking hub, or help launch a sector-specific accelerator. Knowing which level of government controls which levers is key to building effective collaborations.

VC providers represent a fourth axis of collaboration. Over the last decade, CVC has become more common. Companies now co-invest alongside independent venture firms in high-potential IDEs. These CVC investments typically come with strategic value: access to proprietary customer data, specialized production infrastructure, technical insight, and go-to-market channels.

Independent VCs, in turn, benefit from the partnership. Corporate investors offer credibility, vertical expertise, and often faster pilot implementation with industry customers. This mutualism reduces early-stage risk, improves technology validation, and broadens commercialization options.

Still, the integration of CVCs into broader VC ecosystems remains a work in progress. Strategic misalignment is always a risk. Startups must remain independent enough to pivot and grow, while corporate backers must avoid imposing bureaucracy or chasing short-term returns. For this reason, successful investment relationships rely on *strong governance*—clear mandates, exit clauses, transparency in expectations, and co-investment thresholds that ensure equity between strategic and financial goals.

Ultimately, as shown in Figure 6.2, the ability to govern this complex network of actors—from startups and universities to government bodies and VC firms—defines a company's role in the iEcosystem. Hybrid governance mechanisms must be deployed: strategic objectives must be paired with adaptive contracts, and learning loops must be embedded into performance evaluations. Firms must blend equity investments with research grants, licensing deals with co-development projects—each tailored to the level of familiarity they have with the underlying technology and market demand.

By doing so, companies move beyond being gatekeepers of innovation. They become *custodians of the ecosystem*. They enable value creation across boundaries, build bridges between discovery and deployment, and anchor their long-term competitiveness in the health and vibrancy of the system around them.

Structured Governance of Innovation Networks

Effective governance within iEcosystems demands more than occasional partnerships or opportunistic collaborations. It requires a carefully structured orchestration of coordination mechanisms—ones that can address the complexities of knowledge exchange, balance divergent incentives, and bridge gaps in information and trust. In these environments, where diverse players such as corporations, IDEs, universities, public institutions, and venture capitalists converge, traditional governance approaches—whether hierarchy-based or market-driven coordination—prove insufficient. What is needed instead is a more nuanced approach: a system of hybrid governance in which firms operate as gravitational centers, weaving together both formal and informal mechanisms tailored to the distinctive needs of each relationship (Jackson, 2011).

Figure 6.5 distills four pivotal coordination levers that firms can activate to foster and sustain knowledge flows within the iEcosystem. Each lever plays a unique role, yet their combined impact is greater than the sum of their parts.

The first lever is the cultivation of a shared *iEcosystem macroculture*. This goes beyond internal values and instead takes the form of a collective mindset shaped by shared symbols, storytelling, and belief systems. When such a macroculture takes root, it creates a common interpretive

Coordination Mechanism in iEcosystems	Positive Effects on Knowledge Exchange (Corporates vs. iEcosytem Stakeholders)
iEcosystem macroculture	• Promotes expectations alignment • Fosters a common language • Encourages shared information
Stakeholder reputation	• Leads behaviors • Enhances incentives for stakeholder engagement • Nurtures partnership approach
Access to engagement tools (hackathons, accelerators, prize competitions, ...)	• Sets exchange routines • Assists I-Cap vs. E-Cap alignment • Helps IDEs vs. Corporates integration
Rules of engagement (risk capital, CVC, prizes, resource sharing, ...)	• Decrease transactional costs • Increase opportunism costs • Enhance incentives to partnership monitoring

Figure 6.5 Coordination mechanisms and their positive effects on knowledge exchange.

framework—a language that aligns expectations, lowers cognitive barriers, and encourages trust-based information exchange. Within this cultural fabric, participants feel not only invited but intrinsically motivated to collaborate, knowing that shared understanding will govern their interactions across institutional boundaries.

Secondly, *stakeholder reputation* functions as a powerful, informal governance mechanism. In innovation ecosystems, trust often precedes contracts.

Organizations that consistently honor their commitments, deliver on promises, and engage transparently build a reputation that travels fast and wide. This social capital becomes a magnet for collaboration. A strong reputation signals reliability, reduces the perceived risk of engagement, and fosters a climate in which proprietary knowledge can be shared more openly—confident in the expectation that reciprocity and respect will follow.

Beyond the cultural and reputational dimensions, *access to engagement tools* serves as a pragmatic enabler of coordination. Mechanisms such as hackathons, innovation accelerators, and prize-based challenges do more than generate ideas; they establish structured, recurring routines of interaction. These formats connect corporate innovation teams (innovation capabilities or I-Cap) with business players (entrepreneurial capabilities or E-Cap) and bring together IDEs and corporate sponsors within a shared space for experimentation (see section "Leadership for Balancing Innovation and Entrepreneurial Capabilities" in Chapter 1). As shown in Figure 6.5, such tools promote routine-based knowledge sharing, accelerate mutual learning, and help overcome capability gaps between different players, effectively acting as governance instruments in themselves.

Lastly, a robust governance structure is incomplete without clearly defined *rules of engagement*. These rules encompass equity arrangements, intellectual property frameworks, data-sharing protocols, and access to critical resources like capital, infrastructure, or technical support. When such rules are agreed upon early, they reduce transaction costs and ambiguity, diminish the appeal of opportunistic behavior, and embed accountability mechanisms directly into the partnership model. They grant stakeholders the confidence to experiment, knowing their core interests are protected and that shared innovation outcomes will be fairly allocated.

Together, these four coordination mechanisms—macroculture, stakeholder reputation, engagement tools, and rules of engagement—form a governance blueprint that enables firms to operate with the agility of startups while retaining the scale and resilience of mature enterprises. As illustrated in Figure 6.5, their combined effect is transformative: they lower the friction of knowledge exchange, align strategic and operational rhythms between corporates and ecosystem players, and sustain the hybridization of I-Cap and E-Cap over time. Ultimately, this structured yet adaptive governance approach is what empowers innovation ecosystems to support not just one-off projects but enduring, strategic innovation initiatives that evolve and scale across cycles of change.

References

Arora, A., Belenzon, S., Patacconi, A., & Suh, J. (2020). The changing structure of American innovation: some cautionary remarks for economic growth. *Innovation Policy and the Economy, 20(1),* 39–93.

Bettanti, A., Lanati, A., & Missoni, A. (2022). Biopharmaceutical ecosystems: a stakeholder model and the case of Lombardy. *The Journal of Technology Transfer*, *47*, 1948–1973.

Chesbrough, H. (2003). The logic of open innovation: managing intellectual property. *California Management Review, 45(3),* 33–58.

Cohen, S. (2013). What do accelerators do? Insights from incubators and angels. *Innovations: Technology, Governance, Globalization*, *8*(*3*), 19–25.

Denison, D. (1990). Bringing culture to life: The role of organizational culture in strategy and performance. *Organizational Dynamics, 18(2),* 4–19.

Forrest, J. E., & Martin, M. J. (1992). Strategic alliances between large and small research intensive organizations: Experiences in the biotechnology industry. *R&D Management, 22(1),* 041–054.

Frølund, L., Murray, F., & Riedel, M. (2017). Engaging in regional innovation ecosystems: Six questions to get your university partnerships right. Working Paper. *MIT Lab for Innovation Science and Policy, July.*

Gardner & Ibarra (2017). How to Capture Value from Collaboration, Especially If You're Skeptical About It. *Harvard Business Review.*

Hagedoorn, J., & Zobel, A. K. (2015). The role of contracts and intellectual property rights in open innovation. *Technology Analysis & Strategic Management*, *27*(*9*), 1050–1067.

ISO (2019). ISO 56002:2019 – Innovation management system – Guidance. *International Organization for Standardization.*

ISO (2024). ISO 56001:2024 – Innovation management system – Requirements. *International Organization for Standardization.*

Jackson, D. J. (2011). What is an innovation ecosystem. *National Science Foundation, 1(2),* 1–13.

Jaruzelski, B., Loehr, J., & Holman, R. (2011). The Global Innovation 1000: Why Culture Is Key. *Strategy+Business, Issue 65, Booz & Company.*

Keller, M. R., & Block, F. (2013). Explaining the transformation in the US innovation system: the impact of a small government program. *Socio-Economic Review, 11(4),* 629–656.

Makridis, C. A. (2019). Does Culture Pay? Evidence from Crowdsourced Employee Engagement Data. *MIT Sloan School of Management.*

Ries, E. (2011). The Lean Startup: How Today's Entrepreneurs Use Continuous Innovation to Create Radically Successful Businesses. *Crown Currency.*

Schein, E. H. (2010). Organizational Culture and Leadership (Vol. 2). *John Wiley & Sons.*

Slowinski, G., & Sagal, M. W. (2010). Good practices in open innovation. *Research-Technology Management, 53(5),* 38–45.

Spreitzer, G., Garrett, L., & Bacevice, P. (2015). Should your company embrace coworking? *MIT Sloan Management Review, 57(1),* 27.

Watkins, M. D. (2013). What Is Organizational Culture and Why Should We Care? *Harvard Business Review.*

Appendix: ISO 56001 Elements vs. ISO 56000 Principles

The International Organization for Standardization (ISO) 56001 (ISO, 2024) standard represents a significant step forward in the codification of innovation management systems. Building upon the conceptual groundwork established by ISO 56000 (ISO, 2020), this standard moves from broad principles to a structured operational framework—providing organizations with a comprehensive system for managing innovation in a way that consistently generates value.

At its core, ISO 56001 connects the *fundamental principles* of innovation—such as adaptability, systems thinking, and future-focused leadership—with the *practical components* of an innovation management system. As illustrated in Figure A.1, each operational element ranging from planning and support to leadership and performance evaluation is directly mapped to one or more foundational principles from ISO 56000. This alignment is not incidental; it is the very essence of ISO 56001's architecture.

Throughout the main chapters of this book, the ISO 56000 principles have been explored in depth. These principles form the philosophical and strategic core of effective innovation management and shape the mindset required to lead innovation in today's complex environments.

In contrast, this Appendix is dedicated to the ISO 56001 elements—the operational building blocks of an innovation management system. For readers seeking to move toward certification, this section offers a more practical lens, linking each process-oriented component of the standard to its underlying principles. As visualized in Figure A.1, each element is directly connected to specific ISO 56000 principles, creating a coherent, principle-based system architecture.

ISO 56001 Elements	ISO 56000 Principles
Operation	Systems approach
Performance Evaluation	Systems approach
Improvement	Systems approach
	Adaptability
Planning	Adaptability
	Managing uncertainty
	Strategic direction
Support	Exploiting insights
Context of the Organization	Culture
	Realization of value
Leadership	Future-focused leaders

Figure A.1 ISO 56001 Elements vs. ISO 56000 Principles.

This mapping is not merely illustrative—it is instrumental. It highlights how ISO 56001 does not function as a static checklist but as a dynamic system grounded in enduring values. The figure underscores the deep interdependence between what an innovation system is designed to do and the principles that guide how it behaves.

Each of the following sections will examine these ISO 56001 elements individually, providing clarity on their function, purpose, and alignment with ISO 56000 principles. This will equip the reader with both the conceptual understanding and the practical insight needed to design, refine, or assess an innovation management system in line with the ISO 56001 certification standard.

Operation

The Operation component of ISO 56001 is where innovation moves from aspiration to action—it is the engine room of execution. This element captures the practical heartbeat of innovation, turning strategic ambitions into real, measurable outcomes through a wide range of initiatives. These initiatives are not limited to grand breakthroughs or moonshots—they include the everyday activities that accumulate to drive progress. From spotting nascent opportunities, developing early-stage prototypes, to testing bold hypotheses and delivering new offerings to the market, the operational phase is where ideas begin to live and breathe. Executing innovation effectively requires much more than a checklist of tasks. It demands a cohesive system of processes that align innovation efforts with the broader organization. Functions like research and development, marketing, operations, and supply chain must move in concert—not as silos, but as an integrated ecosystem. When these gears are well-oiled and synchronized, innovation efforts build upon one another, multiplying impact over time. This cross-functional cohesion fosters not only collaboration and consistency but also minimizes friction and inefficiencies that often derail promising initiatives. Yet, innovation is not a one-size-fits-all endeavor. Operations must be capable of flexing across different types of innovation—whether incremental improvements that fine-tune existing products, radical changes that reshape core offerings, or disruptive ideas that redefine markets entirely. Furthermore, the system must be adaptable to various innovation models, from top-down strategic initiatives to grassroots, user-driven experimentation, and co-creation with external stakeholders. ISO 56001 recognizes that operational excellence in innovation requires systems thinking. By treating innovation as an interconnected web of relationships and dependencies—rather than isolated projects—organizations can more effectively coordinate departments, streamline processes, and nurture conditions for innovation to flourish. Chapter 5 explores this perspective in depth, showing how systems thinking becomes the backbone of a resilient and performance-oriented innovation management system. It is here, in the operational core, that the foundation is laid for turning innovative potential into sustained value.

Performance Evaluation

In the architecture of ISO 56001, Performance Evaluation plays a vital role—not just as a control mechanism, but as a compass for learning, accountability, and strategic coherence. It provides the organization with a structured way to examine whether its innovation initiatives are delivering what they promise: tangible outcomes, real value, and alignment with long-term strategic ambitions.

At its core, this element establishes a clear framework of criteria, metrics, and feedback channels through which innovation activities can be continuously observed and assessed. It allows teams to monitor progress, ask the right questions—Are we achieving our objectives? Are we generating the expected value? Are we still moving in harmony with our strategic direction?—and make adjustments based on evidence, not assumptions.

What sets this dimension apart is its focus on closing the loop. Performance evaluation isn't just about measurement; it's about learning. The insights that emerge from successes and failures alike are fed back into the system to refine processes, recalibrate goals, and enhance decision-making. In this sense, performance evaluation becomes a living, breathing mechanism—one that helps innovation stay responsive and resilient in the face of uncertainty.

Importantly, this is not a standalone activity. It is rooted in the principle of a systematic approach, which recognizes that performance must be evaluated in context: market dynamics are constantly shifting, stakeholder expectations evolve, and internal priorities adapt. Evaluation must account for these moving parts, acknowledging the interdependencies across teams, initiatives, and external forces.

To do this effectively, organizations can adopt a range of tools—balanced scorecards to map innovation goals against outcomes, dashboards to track progress in real time, or dedicated evaluation committees to review portfolios and ensure they stay strategically relevant. These mechanisms support transparency, encourage honest reflection, and help prevent common pitfalls like project inertia or strategic drift.

Ultimately, when performance evaluation is embedded as a systemic practice, it fuels a virtuous cycle of improvement. It becomes the thread that weaves together insight, adaptation, and long-term value creation.

Chapter 5 returns to this idea, exploring how a systems-thinking lens on performance can transform evaluation from a compliance exercise into a cornerstone of innovation success.

Improvement

In the ISO 56001 framework, Improvement is not a one-off correction, nor merely a procedural refinement. It is the beating heart of continuous evolution within an innovation management system—an ongoing commitment to learning, adapting, and advancing in the face of change.

This element draws directly from the insights gathered through performance evaluations. It is through those analyses that patterns emerge, inefficiencies are uncovered, and opportunities for progress take shape. But improvement goes beyond reacting to what doesn't work; it's about proactively shaping what could work better. It transforms data into direction, and lessons into leverage.

Anchored in the principles of Systemic Approach and Adaptability, improvement demands a wide-lens view of the entire innovation ecosystem. Changes made in one area—whether a process tweak or a shift in resource allocation—inevitably ripple outward. As such, improvement efforts must be coherent and interconnected, ensuring that local optimizations do not compromise overall innovation performance.

At the same time, this element reflects the essence of adaptability: a willingness to question assumptions, embrace uncertainty, and respond fluidly to both internal signals and external shocks. Whether prompted by shifting customer needs, technological breakthroughs, or competitive pressures, the organization must be ready to rethink, restructure, and realign. In practice, improvement can take many forms: post-intervention reviews that analyze what went right or wrong; deep-dive insights that uncover latent potential; kaizen-style routines that embed incremental change in daily work; or benchmarking exercises that compare performance against best-in-class standards. Each mechanism contributes to an organizational culture that values inquiry over inertia. Ultimately, improvement is not just a set of activities—it's a mindset. A mindset that treats every result, whether successful or not, as a steppingstone toward greater capability. It encourages reflection not as a luxury but as a discipline. And it sees disruption not as a threat but as a prompt for reinvention. Chapters 4 and 5 explore how the principles of systemic coherence and adaptive agility intersect to support a robust framework for managing innovation—not only efficiently but also resiliently and sustainably.

Planning

Within the ISO 56001 framework, Planning serves as the vital connective tissue between an organization's long-term innovation vision and the practical steps required to bring that vision to life. It is through this element that strategy becomes action—objectives are clarified, resources are allocated, and innovation portfolios are built and refined.

But ISO 56001 redefines planning as more than a one-time blueprint. It embraces planning as an evolving, adaptive process—one that must remain fluid in the face of change. Rather than locking organizations into rigid paths, it encourages a living plan that grows with market shifts, stakeholder input, and internal developments.

At its core, this approach weaves together the principles of Adaptability, Strategic Direction, and Uncertainty Management. These principles elevate planning from a procedural necessity to a strategic enabler. Planning becomes not only about deciding what to do but also about preparing for how to respond when conditions change—and doing so without compromising the organization's overarching goals.

Scenario planning and strategic forecasting become critical tools in this dynamic process. They allow organizations to explore multiple possible futures, test assumptions, and anticipate disruptions before they happen. Managing uncertainty means recognizing that the future can't be predicted with precision, but it can be prepared for. It involves identifying key variables, assessing their likelihood and potential impact, and building resilience into the plan from the outset.

Another crucial dimension is inclusiveness. Effective planning demands input from across the innovation ecosystem. Engaging diverse stakeholders—from internal departments to external partners—improves both the quality and the legitimacy of decisions. It fosters buy-in, enhances foresight, and surfaces insights that may otherwise be overlooked.

Planning in the ISO 56001 context also acts as the operational bridge between vision and execution. It aligns innovation efforts with strategic intent and ensures that day-to-day actions are pulling in the same direction. Roadmaps, governance structures, and portfolio management tools all contribute to this alignment, creating a shared rhythm across teams and functions.

In this way, planning becomes more than scheduling or forecasting—it becomes a tool for navigating complexity. It balances ambition with realism, direction with flexibility. It enables organizations to remain agile, collaborative, and forward-looking, even in the face of uncertainty.

Chapter 2 delves into the broader theme of stability and uncertainty, setting the stage for Chapter 3's exploration of how these forces are managed within planning activities under a coherent innovation strategy.

Support

In the architecture of ISO 56001, Support functions as the backbone of innovation—an essential element that empowers initiatives to move from concept to reality. Without a robust support structure, even the most promising innovation strategies risk stagnation. This component ensures that the right resources, capabilities, and cultural conditions are in place to nurture innovation consistently and effectively.

At its essence, support is about more than just tools or funding—it's about enabling an environment where creativity, exploration, and continuous learning can thrive. It requires organizations to invest not only in systems but also in people. The development of internal capabilities, the cultivation of cross-functional collaboration, and the integration of data and insights into everyday decision-making all contribute to a thriving innovation ecosystem.

Central to this element is the ISO 56000 principle of exploiting insights. Support means creating mechanisms to gather, distill, and activate knowledge—transforming abstract signals into actionable strategies. Insights may emerge from customer interactions, competitor movements, market trends, or internal reflections. What matters is the organization's ability to make those insights visible and usable across all levels and departments.

Effective support systems include structured talent development programs that empower employees to build innovation-related competencies. They also include technological infrastructure—digital platforms, collaborative tools, and data analytics engines—that connect teams and dissolve traditional silos. These systems ensure that knowledge flows freely and is not stored within isolated departments or hierarchical layers.

When well designed, the support element ensures that innovation is grounded in real-world needs and opportunities. It enables teams to experiment with confidence, make informed decisions, and respond quickly to feedback and change. It also plays a key role in shaping culture, fostering curiosity, resilience, and a willingness to challenge assumptions.

In this sense, support is not simply a backdrop to innovation—it is an active force that shapes its success. It gives form to ambition, structure to experimentation, and coherence to the many moving parts of the innovation process.

Chapter 3 examines how organizations can design and embed support systems that systematically harness insights, unlocking the full potential of innovation across the enterprise.

Context of the Organization

In the ISO 56001 framework, understanding the context of the organization is not merely a preliminary step; it is a foundational act of awareness. It shapes how innovation is envisioned, structured, and ultimately executed. This element invites organizations to step back and consider the broader landscape in which they operate—both internally and externally—before they move forward with bold innovation agendas.

This means taking stock of a wide range of influencing factors: shifting market dynamics, evolving technologies, customer and employee expectations, changing regulatory environments, and the increasing pressure for environmental and social accountability. Only by embracing this comprehensive view can organizations ensure that their innovation efforts are relevant, strategic, and sustainable.

But context is not just about awareness; it is about orientation. It provides the lens through which strategy is shaped and innovation culture is cultivated. The ISO 56000 principle of culture is tightly interwoven with this element, as it highlights the role of openness, trust, and collaboration—both within the organization and across its external ecosystem. A culture of innovation flourishes in environments where institutions, partners, and stakeholders are engaged with transparency and mutual respect.

This outward and inward view allows organizations to align internal strengths with external opportunities and constraints. It encourages them to design innovation systems that not only respond to market signals but also anticipate them, making room for more resilient, adaptive, and future-fit approaches.

Crucially, context is also where the idea of value realization comes into play. Innovation is not an end in itself. It must deliver value—economic, certainly, but also societal and environmental. Understanding the context enables organizations to target their innovation initiatives toward outcomes that matter. That means evaluating industry trends, decoding customer behaviors, navigating competitive forces, and grappling with the broader challenges of sustainability and inclusion.

Organizations that commit to continuous context reassessment are far better positioned to stay ahead of disruption, mitigate emerging risks, and align their innovation systems with long-term stakeholder expectations. This situational intelligence reinforces the adaptability of the innovation management system, while also anchoring it in the human realities of the world it seeks to improve. In this light, context becomes more than a strategic input; it becomes a moral and cultural compass. It ensures that innovation is not detached from the environment but deeply embedded in it, both as a response and as a contribution. Chapter 6 delves into how contextual awareness enhances the business dimension of innovation, while Chapter 3 explores how it informs stakeholder engagement and value realization, particularly when broader societal contributions are at stake.

Leadership

Within the ISO 56001 standard, Leadership is not simply about occupying a position of authority; it is about embodying a purpose, setting a direction, and cultivating the conditions in which innovation can truly take root and thrive. It is leadership that sets the tone, the vision, and the cadence for the entire innovation management system.

To lead innovation is to do far more than approve budgets or green-light projects. It requires the active presence and intentional influence of top management in establishing a clear, resonant vision—one that not only inspires but also aligns resources, people, and priorities across the organization. Innovation cannot remain an abstract ambition; it must be translated into specific objectives, anchored in strategic direction, and sustained through shared commitment.

This element is deeply aligned with the ISO 56000 principle of future-oriented leaders—individuals who look beyond the horizon, anticipate trends, and mobilize their teams to act on what's next. But foresight alone is not enough. These leaders also bring emotional intelligence, resilience, and a genuine commitment to cultural transformation. They are not distant visionaries but active participants in the innovation journey.

True innovation leadership requires the courage to create space for experimentation and learning, even when it means tolerating ambiguity and embracing failure. That means fostering an environment built on psychological safety, where people at all levels feel empowered to voice ideas, challenge assumptions, and explore unconventional paths. Trust and intellectual openness are not soft values; they are strategic enablers of breakthrough thinking.

Moreover, innovation rarely happens in isolation. Future-facing leaders understand the power of networks, and they act as connectors, building bridges with external partners, stakeholders, and ecosystems that can enrich the organization's perspective and amplify its capabilities. They promote collaboration not only within teams but also beyond traditional boundaries.

This leadership is not left to chance—it can and should be nurtured. Organizations can implement deliberate mechanisms such as leadership development programs, peer learning cohorts, and visible role modeling to embed innovation into the leadership DNA. They can institutionalize innovation-supportive behaviors by rewarding curiosity, encouraging exploration, and normalizing strategic risk-taking.

In this way, leadership becomes a distributed force—embedded not just in titles but in the everyday decisions and behaviors that shape the culture. It is leadership that makes innovation a shared mission rather than a siloed initiative, enabling alignment across strategy, structure, and spirit.

Chapter 1 explores in depth the traits, roles, and responsibilities of innovation-driven leaders, examining how they influence ecosystems, mobilize action, and create the cultural foundations on which sustainable innovation can flourish.

Conclusions

This Appendix analysis reveals that ISO 56001 is far more than a structural checklist or procedural guideline; it is a dynamic and evolving framework designed to guide organizations in shaping and sustaining innovation as a core strategic function.

Each element within the standard plays a distinct yet interconnected role, working in harmony with the broader set of principles that underlie the system. This alignment between structure and philosophy is what gives the ISO 56001 framework its transformative potential. Innovation, under this lens, is not treated as a peripheral or experimental activity but as a deeply embedded capability—central to long-term competitiveness and relevance.

Crucially, ISO 56001 does more than encourage innovation—it equips organizations with practical tools and systemic approaches to navigate complexity, manage uncertainty, and respond with agility to shifting market dynamics. Whether through adaptive planning, insight-driven support systems, or resilient leadership practices, the standard enables organizations to remain both grounded and future-facing.

Each element contributes in its own unique way to building an environment where innovation is not only possible but also expected. By weaving together strategic direction, operational execution, cultural openness, and continuous improvement, the framework fosters an ecosystem where innovation can emerge organically, scale sustainably, and generate enduring value.

In this light, ISO 56001 is not merely a compliance tool; it is a blueprint for cultivating innovation as a way of thinking, organizing, and acting. It is a guide for those who seek to embed innovation not as a moment but as a mindset.

References

ISO (2020). ISO 56000:2020 – Innovation management – Fundamentals and vocabulary. *International Organization for Standardization.*

ISO (2024). ISO 56001:2024 – Innovation management system – Requirements. *International Organization for Standardization.*

Index